Barks, Roars and Siren Songs

Also by Michael Bright

The Living World
There Are Giants in the Sea

Barks, Roars and Siren Songs

MICHAEL BRIGHT

A Birch Lane Press Book
Published by Carol Publishing Group

A Birch Lane Press Book
Published by Carol Publishing Group
Birch Lane Press is a registered trademark of Carol Communications, Inc.

Editorial Offices
600 Madison Avenue
New York, NY 10022

Sales & Distribution Offices
120 Enterprise Avenue
Secaucus, NJ 07094

In Canada: Musson Book Company
A division of General Publishing Co. Limited
Don Mills, Ontario M3B 2T6

First published as *The Dolittle Obsession* by Robson Books Ltd, London

Manufactured in the United States of America
10 9 8 7 6 5 4 3 2 1

Carol Publishing Group books are available at special discounts for bulk purchases, for sales promotions, fund raising, or educational purposes. Special editions can also be created to specifications. For details contact: Special Sales Department, Carol Publishing Group, 120 Enterprise Ave., Secaucus, NJ 07094

Library of Congress Cataloging-in-Publication Data

Bright, Michael.
Barks, roars and siren songs : how animals talk to us and how we talk back / by Michael Bright.
p. cm.
"A Birch Lane Press book."
Includes bibliographical references and index.
ISBN 1-55972-086-7
1. Animal communication. 2. Human-animal communication.
I. Title.
QL776.B754 1991
591.59—dc20

91-27046
CIP

Contents

Introduction and Acknowledgements vii

1 Talking to Animals 1

2 Apes and Man 16

3 Conversations in the Wild 79

4 Dolphins and Man 140

5 Animals and Language 216

Bibliography 226

Index 228

Introduction and Acknowledgements

Can captive chimpanzees or gorillas have a meaningful conversation with people by making signs with their hands? Can we translate the whistles and squeaks made by a dolphin into English with the help of a computer? What are humpback whales communicating in their long and sonorous songs? What does a dog really mean by jumping up and licking your face? Can a horse or a cormorant count? Why should the military show an interest in man-animal communication research? Can you really ride a wild dolphin? Do animals have languages of their own and can they 'think'?

Barks, Roars and Siren Songs is fundamentally about people and animals and ways in which the former have contrived to 'talk' with the latter. It explores an area of scientific endeavour which touches on animal intelligence, thinking, self-awareness, language, animal communication, and a whole host of other subjects that present a veritable minefield of interpretive problems for the unwary. As an informed and interested outsider but by no means an 'expert', I have tried to find my way through the accumulated knowledge and the often heated and bitter arguments as best I could to present a general picture of the state of man-animal communication.

Many people have given me guidance and information along the way to preparing this book. Those I would like to acknowledge for their help include (in alphabetical order):

Jeffrey Boswall, Tim Clutton-Brock, Tom Eisner, John Ford, James Gould, Peter Greig-Smith, John Gribbin, Donald Griffin, Tony Hawkins, David Helton, Louis Herman, Eric Hoyt, John Krebs, Alison Jolly, Roger Jones, Peter Marler, Maura Mitchell, Keiran Mulvaney, Ken Norris, Katherine Payne, Roger Payne, James Serpell, Robert Seyfarth, Peter Slater, Paul Spong, Herb Terrace, Jeremy Thomas, Peter Tyack, Haven Wiley, Forrest G Wood, and the directors of the Dolphin Research Center (Florida).

To them and all the dedicated researchers, who are too numerous to mention here but most of whom are credited in the text, I offer my grateful thanks.

The bibliography will lead the reader to more comprehensive and finely-focussed works which contain reference lists of relevant scientific papers.

1

Talking to Animals

The Dream

'How marvellous it would be,' suggested Professor Herb Terrace of Columbia University (New York), 'if we could teach an animal a sign language and then release it into the wild where it could act as an interpreter with others of its kind. Think of the insights we could gain into animal behaviour.'

It was a mind-boggling thought and it certainly caught *my* imagination. Wouldn't it be amazing, I thought, to be able to talk to a representative of another species? Imagine the things we might discover about the world in which we live and our place in nature seen through that animal's eyes. Is the notion too far-fetched or might we be able, some day, to become modern-day Dr Dolittles and talk to the animals?

The very concept is fraught with difficulties. If we are to teach an animal anything, it must understand what we are teaching it. Animals, however, are not necessarily on our wave-length. After all, many animals 'talk' using channels of communication which we infrequently use: ants communicate with smells and vibrations; blind mole-rats bang their head against the tunnel roof; wolves have a complex body-language; fiddler and ghost crabs wave their pincers in the air; fireflies flash light at each other; songbirds sing with trills so rapid that they cannot be analysed by the human brain; bats, crickets and dolphins emit sounds far above the frequencies that we can detect, while giant whales and elephants communicate at frequencies too low. It's one thing to come up with some technological breakthrough that could convert our signals – by slowing them down, speeding them up, stretching them, or compressing them – into signals which could be understood by a trained captive animal, but quite another to get it to be accepted by others of its own kind after it had been returned to the wild.

Pet Talk

Sign language aside, there are precedents for humans talking to animals. Many creatures, including those we have about us during our everyday lives, stimulate that need to communicate with other species. Many of us talk to and seek affection from pets, whether they be cats, dogs, parrots, budgerigars, hamsters, rabbits or white mice. And there is evidence that animals listen and respond to the utterances that we make, even if they do not fully appreciate what is being said.

Shepherds are able to send their sheepdogs this way and that by means of a simple whistle language. Equestrians ease their charges through dressage and jumping with a subtle combination of touch and verbal encouragement. But do the animals really 'understand' what is being said to them, or do they simply respond automatically with a view to some future reward, whether it be a tasty morsel or a pat on the back?

We often *think* they understand. How often have you heard a pet-owner say 'my cat understands my every word' or 'my dog knows what I'm going to do even before I do it'? In a recent survey, mentioned in Bruce Fogle's entertaining book *Pets and their People*, it was found that nine out of 10 pet owners believe their pets understand their moods. But do the animals actually realize how we are feeling or know what we are saying? And can they perceive, by some mind-reading process, our unspoken thoughts.

There are simple alternative explanations. For instance, a move towards a dog's leash will tell the pet that it is time for a walk. But dogs are capable of picking up much less obvious clues. Most animals, for example, have a remarkable appreciation of the passage of time. The move, at the same time each evening, to put on a coat prior to a walk is anticipated by the animal. To the uninitiated, it could look as if the animal had read our mind. In reality, its internal biological clock had simply told it that the time had come for that particular activity.

In nature there are some amazing feats of timing. The honey bee, for example, is guided by some biological clock to go to particular species of flowers only at certain times of the day. The visit coincides with a periodic production of nectar.

Animals in zoos show this very well. Big cats pace up and down their cage or enclosure and seals and sea lions swim about frantically just before mealtimes. Interspersed are short periods in which they stare intently in the direction from which the keeper will come. They appear to become increasingly agitated until, at the crescendo of activity, the keeper arrives with the food. Likewise, in the wild, animals have feeding routines. An appreciation of the passage of time is a survival mechanism. It ensures, for example, that day-adapted creatures are out of harm's way before the night-adapted creatures arrive. Curiously, humans show this ability to sense the time accurately when they are asleep. How often, when catching an early train or plane, have you set your alarm clock, only to wake up just before it goes off?

As far as your 'clairvoyant' pet is concerned, there may be even more subtle cues that warn of walks. Lyall Watson, in his book *Supernature*, considered the fact that we emit all sorts of radiations, ranging from heat to the electrical activity of our muscles. These may reflect changes in our emotional state that can be detected by our pet. The simple anticipation of a walk on our part might result in the emission of radiations that indicate to an animal that something is about to happen.

Then there is body language – unconscious changes or movements of the body that transmit messages other humans unknowingly pick up. In an often-quoted series of experiments at the University of Chicago, Eckhard Hess noticed that if somebody was interested in something, the pupils in their eyes became larger, whereas if they looked at something uninteresting or distasteful the pupils contracted. The culmination of the experiment came when Hess presented male test subjects with two pictures of an attractive girl. In one picture, her pupils were touched up to look larger. The men were asked to say which picture they preferred. Most said they could see no difference, but their own eyes gave them away. Hess monitored the pupil size of the men as they viewed the two pictures. Their pupils widened in response to the girl's wide-eyed look. They had unconsciously recognized that her eyes were saying 'I am interested in you'.

Interestingly, cats' eyes do the same. When a cat is presented with something it likes, its slit-like pupils will dilate. The signal can be confusing, though, for a cat's pupils will also dilate if it is frightened. In other words, it uses the same eye movement for events to which its emotional responses are quite different. It is as if the creature has been stimulated to such an extent – either by pleasure or fear – that it needs to obtain more information; hence the eyes open wider. The movement could also act as a signal to others – either friend or foe – which says, 'I am motivated to react to you'. Humans, though, can lie. They can put out contradictory signals. A person can be using a vocal language to say 'I am calm', while their body language, such as sweating palms or brow, reveals them to be very excited or agitated indeed.

Could it be that animals, like our fellow human beings, can detect body changes of which we ourselves are not consciously aware? Animals, after all, are known to have some of their senses developed to an extent that they are far superior to our own. Dogs and deer can detect infinitesimally small amounts of smells, and some birds, such as vultures and albatrosses, can smell a carcass at a range of many kilometres. Owls, sitting high in a tree or hovering overhead, can hear and locate the almost imperceptible movement of a mouse in the grass. Bats, oil birds and cave swiftlets can use sound in order to fly safely through caves in complete darkness. Many animals are reputed to be able to detect minute vibrations and predict earthquakes. Sharks, rays and the duck-billed platypus can detect minute electrical currents emitted by the muscles of their prey animals and so can find a meal even when it is buried below sand or mud.

It is this very ability of animals to detect small movements or changes in bodily functions that confuse our interpretation of their understanding or intention. We unconsciously cue animals to behave in a certain way and then conclude they have initiated the behaviour themselves. This is one of the fundamental problems facing all those involved in human-animal communication research, and it is exemplified by the classic story of Clever Hans.

Clever Hans

Clever Hans was a Russian stallion bought by a retired German schoolteacher, Wilhelm von Osten, in 1900. Osten was convinced that animals possess an intelligence equal to that of man, and had spent much of his spare time trying to prove the point. He started with a rather troublesome bear and a dim-witted cart horse, both of which showed not the slightest inclination to fill their heads with multiplication, division, addition or subtraction, let alone square and cube roots.

Clever Hans, though, was of a different calibre altogether. Osten taught the horse the numbers one to nine with the help of a row of skittles, then replaced the skittles with numbers on a blackboard. In a short time Clever Hans could do quite complex calculations, including square roots and, after an initial test conducted by an eminent mathematician, was considered to have demonstrated the mathematical capability of 'an intelligent 14-year-old schoolboy'.

Clever Hans enjoyed world acclaim, but the scientific community was sceptical and submitted the wonder-horse to five weeks of strict tests. Each time the horse was given a calculation to solve, the sum was drawn on the blackboard and Clever Hans tapped his hoof on the ground to give the answer. Scientific observers were astonished as Clever Hans kept on getting the sums right. They could find no obvious signs of trickery.

Then one of them, Oskar Pfungst, a psychologist from Berlin, had a brainwave. Could the horse get the correct answers if nobody in the stable could see the calculation? Everybody was moved behind the blackboard so only the horse could see the numbers. Thereafter, he was consistently wrong. What the horse had been doing, proposed Pfungst, was to take 'cues', either from his trainer or from other people in the room. He was able to spot very subtle movements, such as a minute nod with a movement of just one fifth of a millimetre, and detect the tension experienced by everybody as he reached the correct number. Clever Hans, concluded Pfungst, was very perceptive but not actually as bright as his name suggested.

There had been an indication that this was likely in 1891, when a horse and a dog that performed on the stage of London's Royal Aquarium were known to be covertly cued by their trainer. Five years later, at about the time Clever Hans was astonishing the western world, the learned journal *Nature* published an article by Joseph Meehan which revealed the methods employed. The horse counted out with its hoof, just as Clever Hans had done, and it was stopped at the correct answer by the trainer suddenly looking away from the horse. The dog was trained to respond to a slight movement of the trainer's hand.

Osten was so humiliated by Pfungst's revelations and the universal scientific support the psychologist had won that he died a broken man. Hans, however, lived on. Osten had left him in his will to a German businessman Karl Krall. What followed was even more remarkable, and it is curious that this later period is often omitted from more recent accounts of the Clever Hans story.

Clever Hans was joined in his endeavours by two Arab stallions with the appropriate names of Muhamed and Zarif, a Shetland pony called Hanschen, and a blind horse called Berto. Within weeks the newcomers had learned how to subtract and add. Later, according to Krall and reported in his book *Dikende Tiere,* Diving Animals, they went on to work out difficult calculations, to spell, and to use a sign language.

Krall invited distinguished professors to come and test the horses. He made himself scarce on test days, allowing the mathematicians and psychologists to work alone with the test animals. In order that cueing was eliminated completely, the scientists stood behind screens and observed the horses through peep-holes. According to some reports, such as that by Professor Mackenzie of Genoa, Professor Claparede of Geneva (who had already been taken in by Clever Hans's earlier act), and Professor Assagiola of Florence, the horses were able to work out quite difficult calculations such as the square roots of long numbers, and almost always came up with the correct answers. There were even reports of the horses 'talking' in their hoof-tapping sign language. Zarif, so it was said, stopped work and indicated he was tired. His leg was hurting! And when asked in the same sign language why he did not speak, rather than tapping his foot, Muhamed replied that he did not have a voice.

What is to be made of these reports, which were written in 1912-13, is difficult to say. Are we to take the observations and the steps taken to avoid cueing to be accurately recorded? Has anybody tried to repeat the experiments? There have been a few tests with dogs, whose owners have followed the same quietly spoken, persuasive teaching methods adopted by Krall. In fact, there was a spate of these kinds of animal performances in Germany during the 1900s; but nobody has really taken them seriously. As for the horses themselves, they were taken to the front during World War I and were killed in action.

There was, however, one more horse worthy of mention – Lady Wonder, the talking horse of Richmond, Virginia. In their book *Living Legends* John Michell and Robert Rickard gathered together newspaper and magazine stories about Lady, as she was known locally, and found that not only did the horse communicate with people by prodding a keyboard with her muzzle, but also, it was claimed, she could read minds.

Lady became the local oracle and people from Richmond and neighbouring Petersburg would come and have their fortunes told. They would even ask the horse to predict changes in the weather and the winners of horse races. The police were reputed to have asked her to help them locate missing people. The parapsychologist J B Rhine of Duke University (Durham, North Carolina) carried out some tests with Lady and reached the conclusion, in his *New World of the Mind*, that she really could read minds. In one report in *Fate* magazine in 1963, Jack Woodford recalled one of his first encounters with Lady. He asked the horse to take a guess at his name. The horse tapped out 'Jack'. Then he lied to the horse and told her that his name was not 'Jack'. Lady tried again and tapped out 'Josh', which was the name the Woodfords' grandmother had called him but that nobody else present would have known.

'Motherese'

Steering clear of mind-reading, telepathy and the like, which many pet-owners claim they have experienced with their pets, there are ways in which pet animals *do* 'talk' to us; but these ways are unique to them. It is likely, however, that we misinterpret their behaviour. We place our own feelings on the pet, often ignoring the real reason for the communication.

We blur reality by talking to our pets not in a normal adult manner but in a tone of voice reminiscent of our way of talking to a child. Pets are treated as members of the family and, indeed, are spoken to as if they were humans, not animals. Linguists at the University of Pennsylvania have dubbed the way mothers talk to their children as 'motherese', and this is just the way we talk to pets. Bruce Fogle reminds us that an often-quoted reason for having pets is 'for the children', while in childless families some pets play the role of surrogate offspring. But, of course, they are *not* children – most are adult animals with natural animal instincts.

Much of the way a pet behaves in the home is a throwback to its natural ancestral behaviour in the wild. Dogs, for instance, are domesticated wolves and, as such, go about their everyday lives like wolves; and wolves have an extensive body language. When a dog rolls on its back in front of you, it is demonstrating that it is subordinate to you. You are the 'pack leader' and when you return to the home it is saying to you that it is glad the pack leader has returned; it feels, in fact, more secure now that you are there. Even after a five-minute trip to the papershop, the greeting may be as enthusiastic as if you had been away for five years.

Often your dog will jump up to lick your face. This is reminiscent of the way young wolves lick at the face of adults returning from a kill. It is the way they get the adult wolves to regurgitate food. The behaviour continues into adulthood when the subordinates in the pack greet more dominant animals.

Your dominance is reinforced when you stare briefly at your dog, although a lengthy stare will cause the animal to become frightened and it may bite. The dog will avert its eyes and lower its ears and tail – all signs that it is subordinate.

It is easy to teach a dog to 'shake paws' in the manner of a human greeting, because even this has a throwback to the wild. For the dog it is much more than a greeting or a performance to amuse the neighbours. If a wolf raises a paw to another wolf it is indicating, once again, that it is subordinate. On the other hand, a dog that jumps up at you with ears and tail erect is telling you that it is dominant. Often these are the more troublesome dogs.

Then there are cats that whine at the back door to be let in. They are, in fact, saying a lot more than 'please let me in'. Cats are territorial animals and are happy only if they can briefly visit the boundaries of their patch. A cat faced with a door without a two-way cat-flap is uncomfortable. It cannot carry out its instinctive need to walk the boundary. The whine, in reality, is indicative of a very frustrated cat.

Fellow

Although there have been many claims that pets fully understand what their owners say to them, there has been only one scientific study that indicated there might be some truth in such claims. Fellow was a famous German shepherd dog in the 1920s. He had been in several movies, incuding *Chief of the Pack*, and his owner asserted that the dog could understand about 400 words of speech. The matter was investigated by two psychologists.

Fellow's owner talked to the dog in a quiet conversational manner and never scolded him. The first tests were carried out with Fellow and the owner in sight of each other. The psychologists were impressed with the way in which the dog appeared to understand what it was asked to do. Then man and dog were re-tested, this time with screens between them. Again Fellow did well and responded to 53 commands, such as simple requests to 'sit' and 'roll over', and more difficult ones like 'put your head close to the lady'.

During the same tests Fellow was asked to retrieve objects. He did quite well, but made several errors. He would go to fetch an object, but sometimes it did not match the one described by his owner. What was interesting, though, was the nature of his errors. Sixty years after the event, psychologist Geoffrey Beattie looked at the original report in the *Quarterly Review of Biology*, and in the last of a series of articles in *The Guardian* in 1987 he reconsidered its results.

Fellow had a couple of teeth missing from one side and he found it difficult to pick up objects that were flat on the floor. So, when given the command to fetch the package, he carried out the task successfully if it was standing upright, but avoided it and picked up something else if it was lying flat. Likewise, if a brush was lying with the bristles upwards and therefore liable to spike him in the nose he declined to fetch it; but if the bristles faced downwards he complied with the command. It seemed clear to Beattie that Fellow not only understood the words of his master, but showed he possessed a kind of intelligence that had hitherto been denied to dogs.

Intelligence has been likened by some researchers to flexibility, an ability to change behaviour in response to changes in the environment. Fellow showed flexibility. He had acquired a detailed image of his world based on his experiences. That accumulated knowledge gave him a view of reality. He was then able to incorporate new knowledge and to behave appropriately to any changes in reality. In other words, he demonstrated some degree of 'intelligence'.

The Research

The distinguished Swiss zoologist, Heini Hediger, of the University of Zurich, once told a New York 'Clever Hans' conference of linguists, psychologists, philosophers and those involved in man-animal communication that their work was motivated by 'the age-old burning desire of mankind to take up language contact with other animals'.

But the desire to communicate with animals is not such a frivolous venture as it may seem at first. There is the view, held by some philosophers, that language is prior to thought.

Without language there can be no thinking. So, if animals *do* think, they must have some linguistic means to do it. Access to the mind of a non-human animal, as with the human animal, must be through language. There are two ways in which this may be approached – what I have termed the 'arrogant' way and the 'sympathetic' way.

The arrogant approach is to try to teach animals a man-made language, like that proposed at the outset by Herb Terrace, whether it be spoken English or a gestural language such as American Sign Language for the deaf. This approach, of course, has fundamental problems of usage and interpretation. Would an animal, for example, be able to appreciate and master the grammar of the language? We know what *we* mean by using certain signs in a particular order, but would we know what an animal meant if it used those signs, or even if it understood at all?

The sympathetic way is to gain an understanding of an animal's own utterances or body movements and try to interpret what is being said. Whether the animal is using a 'language' is debatable, as we will see later; but at least the animal would, hopefully, be telling its tale in its own 'words'. This approach also has problems, the basic one being that it would be easy to anthropomorphize.

Both ways are open to other misinterpretations. In any attempt to communicate with another species we could have little understanding of the nature or context of their own utterances. Animals may view the world very differently from the way in which we do. By imposing our man-made languages on captive individuals or by interpreting observations in the wild in human terms, we may obtain not so much information about other species as reflections of ourselves.

We can take a leaf out of the secret books of magicians, illusionists and circus performers. Often their tricks depend on what Paul Buissac calls the 'dialectic of misunderstanding', the way we are hoodwinked by suggestion and expectation. There is an old circus trick known as the 'Kiss of

Death', recalled by Buissac at the New York conference, which illustrates the point. A girl, clothed in a provocative way, is strapped to a bed. A large captive bear is released and it heads straight for the bed. The audience becomes concerned for the girl, expecting to witness either disturbing scenes of ravishment and bestiality, or, at the least, some kind of attack. Instead, the bear appears to kiss the girl and then shuffles back to the cage. The bear's thoughts, unlike those of the audience, were not on the prospect of an obscene performance, but on the carrot it knows is hidden in the girl's mouth.

A study of police dog behaviour also illustrates the way people can be misled by faulty interpretations, and how human behaviour can be substituted for animal behaviour. A police dog is supposed to be able to follow a scent trail and lead its human handler straight to the person who has left the trail. But, in one study, researchers discovered that the police dog-handler was giving the dog subtle clues about his own expectation of the route they should follow. The policeman had unwittingly steered the dog in the direction he thought it should go. The dog was not leading the man: the man was leading the dog.

So, bearing in mind the pit-falls, such as unintentional cueing or anthropomorphic interpretation, that lie before an eager researcher, which animals might best lend themselves to study? Which creature, for instance, is likely to have the brain capacity and the motivation to learn, say, a signing system invented by people? Or, even better, with which animal might we be able to actually talk?

Should it be one of our nearest relatives, the great apes – a chimpanzee or a gorilla? Should it be a dolphin or a whale? Should it be a dog, a cat or even that highly intelligent invertebrate, the octopus? Veteran human-dolphin communication researcher John Lilly defined his criteria for selection in a paper entitled 'Interspecies Communication' in the *Year Book of Science and Technology, 1962*:

A mammalian brain above a certain critical weight, a brain of

> a certain degree of complexity, an adequate vocalization apparatus, a naturally cooperative attitude toward man, sufficient control over emotional impulses (such as aggression, sexual activities, and so on) and ability to learn quickly to select the appropriate sonic and other patterns from the environment.

Lilly's criteria, as we will see later, excludes the apes on the grounds of poor vocal abilities, and also because some can become quite violent when upset. Man's best friend, the dog, fails on most counts. Lilly, of course, was championing the dolphin. But I would guess he and the rest of the scientific community were dumbfounded when Professor Irene Pepperberg, now at Northwestern University (Evanston, Illinois), introduced the animal-man communication world to Alex. Alex is not a dolphin or an ape, nor even a humble monkey or dog. Alex is an African grey parrot.

It is not surprising to find a parrot that can learn words in English – or Russian, French, or Mandarin for that matter – and, indeed, Alex has, over the past 12 years, acquired a vocabulary of 80 English words, all of which are the names of toys, shapes or colours. The remarkable thing about Alex, however, is that he can not only associate these words with objects or the qualities they describe, but also can link together two or more words to convey a new meaning.

Nowadays, in certain scientific circles, parrots are considered 'honorary primates' in terms of learning ability. Until recently, they were considered to be little more than able mimics; although in ancient times the Greek court was once impressed by the way an Indian parakeet could speak a human tongue, albeit in a language that the Greek elders could not understand.

Before Alex's arrival on the communication scene, a parrot was not expected to be able to distinguish between say, colours and shapes or 'same' and 'different'. Yet Alex learned to distinguish between, for example, a blue key and a green ball or a green pen and a blade of grass; and, what is more, he says so. The remarkable Alex can associate adjectives with nouns. Not only that, he can count. Alex can appreciate quantity – up to six of anything – as well as quality. He also understands the concept of 'none'. In other words, he can appreciate the absence of something.

This ability to pick up and seemingly understand certain components of human language does not surprise some researchers. Many species of parrots in the wild are social animals, like many species of primates. And social animals inevitably have a strong interest in communication. James Serpell, of the Companion Animal Research Group at Cambridge University studied Loriine parrots from the Indo-Pacific region. He found that they live in large flocks, but that adult flock members pair for life. As a consequence, they indulge in a large number of ritual displays aimed at keeping potential interlopers at a distance, reinforcing the pair bond, or dissipating aggressive behaviour between the pair themselves. Charles Munn, of Wild Life Conservation International, came to similar conclusions after observing parrots in the rain forests of Peru. He points out that parrots have comparatively large brains for their body size and consequently have even more communicative skills to show us.

Alex, it seems, is simply carrying on his ancestral tradition; but, having been deprived of his native grey parrot 'language', he has learned to use whatever is available. But he has taken it all a stage further. He has learned more than just to describe objects put before him. He can express his own feelings. Irene Pepperberg has drawn parallels between Alex and a two-and-a-half-year-old child. And, like a human child that does not want to cooperate or is bored with what is going on, Alex has discovered the power of the word 'no!'

2

Apes and Man

Close Relatives

In 1923 the distinguished primatologist Robert Yerkes bought a young male chimpanzee from an animal dealer in New York and took him to his experimental facility in Florida. Yerkes was assured by the dealer that the animal had been taken in the eastern part of what was then the Belgian Congo. But this proved to be no ordinary chimp. He was very bright, imitated human actions, and was unlike any of the chimpanzees in the collection. He was, Yerkes thought, an unusual animal, and he gave him the name 'Prince Chim'.

Indeed, Prince Chim made such an impression on Yerkes (albeit a short one: he died of a respiratory disease a year after his arrival) that the scientist mentioned him in his pioneering ape-behaviour book *Almost Human*. He wrote:

> Intelligence is a much-abused word, but it is more comprehensive than alertness, and I am going to risk using it. Everything seems to indicate that Chim was extremely intelligent. His surprising alertness and interest in things about him bore fruit in action, for he was constantly imitating the acts of his human companions and testing all objects. He rapidly profited by his experiences. To say that he was unique among anthropoid apes, or even among chimpanzees, would be rash on a few months' acquaintance

> and in the light of intimate knowledge of only a few of his kind, yet I do know with certainty that hc, an infant ape, was remarkable alike for his observational ability, his varied methods and quickness of learning, and above all for his delight in 'acting'.

Curiously, Chim looked quite unlike his companion, another chimpanzee named Panzee. And as far as their behaviour was concerned they were 'opposites'. When Yerkes took Chim to the Quinta Palatino estate near Havana, Cuba, where Rosalia Abreu kept a private collection of apes, Yerkes's collaborator Professor Harold Bingham drew attention to the fact that Chim, unlike other chimpanzees, was 'interested in the causes or condition of things'.

Chim's appearance was also noted. All the other chimpanzees at Quinto Palatino had white faces, large floppy ears, and coarse, thin hair, but Prince Chim had a small black face, a conspicuous nose, small ears and a thick coat of fine, black hair. He was altogether different from other chimpanzees.

A few years later, another American zoologist, Harold Coolidge, was browsing through the museums of the world in the course of classifying gorillas when in the Musée Royale de l'Afrique in Belgium, he came across a skull which was labelled 'young chimpanzee'. The bones in the cranium, though, were fused, indicating the skull was not from a youngster but from an adult. The skull was smaller than the common chimpanzee skull. Coolidge was intrigued and, after completing his gorilla work, he hunted the museums for more chimpanzee material. Eventually, he revealed, in a paper in 1933, that the skull, and others of its kind, belonged to a 'pygmy' chimpanzee, to which he gave the scientific name *Pan paniscus*.

Coolidge went further. He speculated that the pygmy chimpanzee was the closest living relative to the common ancestor of chimpanzees and man. As John and Mary

Gribbin remark in their book *The One Per Cent Advantage*, Coolidge had proposed 'not so much a missing link as a *living* link with our past'. And one of these living links, it turned out, was the human imitator Prince Chim.

For the next element in the story we must jump a couple of decades. In 1953 Francis Crick and James Watson unravelled the structure of DNA – the 'stuff of life' in which the genetic code of living things is set. This gave the scientists a way of looking not just at bones and skins to work out relationships between species, but at the chemicals that make up living bodies. In recent years, it has been shown that the DNA in the nucleus of the cells and the amino acids in certain proteins enable us to make detailed comparisons of different species. We can take samples from superficially similar animals, such as a rat and a mouse, and not only determine the genetic and amino acid differences between the two but also estimate how long ago the two animals shared a common ancestor. Small changes in the amino acid structure of a protein, for example, seem to take place with some regularity, giving us a means by which to measure evolutionary change. This has been called the 'molecular clock'. The greater the differences between the genes or amino acids in the samples of the species being compared, the further back we have to look for their common ancestor.

When such tests are carried out for apes and man, we find that gorillas, orang-utans, and chimpanzees are all closely related to man. Chimpanzees, however, are closer than the others; and pygmy chimpanzees are the closest of all. Indeed, we share about 99 per cent of our genetic make-up and the amino acids in our proteins with the pygmy chimpanzee.

Even more intriguing is the estimate that our common ancestor with the apes – most likely a pygmy chimp-like creature – lived about five million years ago. All this emerged in a paper published in 1968 by the molecular-clock pioneer Vincent Sarich, of the University of California. Ten years later one of Sarich's followers, Adrienne Zihlman, studied the comparative anatomy of pygmy chimps, common chimps and humans and published a paper in the learned journal *Nature* speculating on the possibility that the pygmy chimpanzee could be used as a prototype for the common ancestor of man and ape.

In the meantime, bone-hunters were uncovering significant prehuman fossil finds in Ethiopia. Don Johanson and Terry White unearthed fragments of 'Lucy', a hominid which lived in East Africa over three million years ago. The specimen was considered to be close to the common ancestral stock which led, on the one hand, to the living apes and, on the other, to man. Zihlman compared the bones of the pygmy chimp with those of 'Lucy', and when the two skeletons were drawn alongside each other, and to the same scale, the match was remarkable. The research teams concluded that the earliest known man-like apes, living three-and-a-half million years ago, were but a stone's throw anatomically, physiologically and intellectually from a small ape-like creature resembling the modern pygmy chimpanzee.

Other researchers, in the fields of psychology, psycholinguistics and animal communication, identified another scientific goal. By studying non-human primate behaviour – particularly that of orang-utans, gorillas and chimpanzees – we could, perhaps, gain some understanding of the intelligence and behaviour of our direct ancestors. And what if we could actually *talk* to an ape, such as the pygmy chimpanzee, and encourage it not only to tell us about the workings of its own mind, but also to reveal something about our intellectual past and inheritance?

At London's Zoological Gardens in the late nineteenth century, a zoo keeper had unwittingly focussed the attention of naturalists on the prospects of talking to a chimpanzee. He was observed by the English naturalist George Romanes. In his book *Mental Evolution in Animals*, published in 1883, Romanes wrote:

> This ape has learned from her keeper the meaning of so many

> words and phrases, that in this respect she resembles a child shortly before it begins to talk. Moreover, it is not only particular words and particular phrases which she has thus learned to understand; but she also understands, to a large extent, the combination of these words and phrases in sentences, so that the keeper is able to explain to the animal what it is he requests her to do.

At about the same time, R L Garner, the American explorer of the 1890s, was teaching a chimpanzee called Moses to 'talk'. Bribed with corned beef, the chimp was taught several words from different languages – 'mama' from English, 'feu' meaning fire from French, 'wie' (how) from German, and the Nkami word for mother 'nkgwe'. After a frustrating few weeks Moses eventually began to respond to training. In *Gorillas and Chimpanzees*, published in 1896, Garner reported his progress:

> In his attempt to say 'mama' he worked his lips without making any sound, although he tried to do so.... With 'feu' he succeeded fairly well, except that the consonant element, as he uttered it, resembled 'v' more than 'f'.... In his efforts to pronounce 'wie' he always gave the vowel element like German 'u' with the umlaut, but the 'w' was more like the English....

Unfortunately, Garner was not trained as a psychologist or any other scientifically acceptable discipline and so his results were considered to be flawed, although some of his observations were verified by later researchers. Since the turn of the century, however, the scientific community has devoted a great deal of time and ingenuity to the subject of ape-human communication, and has taken increasingly serious interest in what chimpanzees and the other anthropoid apes can say and do.

Do You Speak English?

In the first decade of this century, Lightner Witmer was passing Keiths Theatre in Philadelphia and noticed that one of the acts was billed as 'a monkey who made a man of himself'. This aroused Witmer's curiosity and he went in to watch. The act in question was Peter, a 4-6-year-old chimpanzee owned by a man named McArdles. Mr McArdles had trained Peter for two and a half years to perform human-like tricks, such as skating. Witmer decided to test Peter to find out just how far his intelligence extended. In the autumn of 1909 arrangements were made for Peter to visit the Psychological Clinic in Philadelphia. At that time, psychology was developing a more scientific approach to experiments, developments and discoveries, and the tests with Peter were taken very seriously.

Peter was tested for co-ordination and reasoning. He could bang nails into wood, open locks, and retrieve articles from inside boxes. He could imitate the movements of writing, but could not actually write anything. Witmer wrote in *The Psychological Clinic*:

> If Peter had a human face and were brought to me as a backward child and this child responded to my tests as credibly as Peter did, I should unhesitatingly say that I could teach him to speak, to write and to read within a year's time. Peter has not a human form, and what limitations his ape's brain may disclose after persistent effort to educate him, it is impossible to foretell. His behaviour, however, is sufficiently intelligent to make this educational experiment well worth the expenditure of time and effort.

Peter could also say one word: 'mama'. The chimpanzee did not find it easy, articulating while breathing in rather than out, and was reluctant to make the sounds at all. The first consonant 'm' sounded good, but the second 'ma' was more like 'ah' and was difficult to hear. Witmer, however, trained Peter in just a few minutes to make the 'p' sound. He wrote:

> If a child without language were brought to me and on the first trial had learned to articulate the sound 'p' as readily as Peter did, I should express the opinion that he could be taught most of the elements of articulate language within six months' time.

Witmer was also convinced that Peter understood English and went on to make some revolutionary proposals, including the following prediction:

> Within a few years, chimpanzees will be taken early in life and subjected for purposes of scientific investigation to a course of procedure more closely resembling that which is accorded the human child.

Witmer went on to suggest that Peter could first be taught the use of symbols as representing objects – an idea that stemmed from the work of Anne Mansfield Sullivan with Helen Keller.

Helen Keller had been stricken with a disease at the age of nineteen months and became deaf and blind. In 1887 the inventor Alexander Graham Bell recommended to her parents that she be taken to the Perkins School of the Blind in Boston, where Miss Sullivan was working. The two built up a very close relationship. Miss Sullivan, for instance, devised a way in which Helen could communicate with the outside world by pressing the manual alphabet into Helen's hands. In this way she was able to learn the names of objects, and within two years Helen could read and write in braille.

Helen Keller and Miss Sullivan were two remarkable people. Helen went to Radcliffe College, where Miss Sullivan signed the lectures into Helen's hand. Helen graduated in 1904 and went on to bring public attention to the plight of people suffering the same affliction. Helen's lectures were voiced by a translator, although by now she could make audible and understandable sounds. She had learned to speak by placing her hands on Miss Sullivan's larynx, where she could sense the vibrations.

With Helen Keller's example in mind, Witmer suggested that Peter the chimpanzee could first learn to appreciate symbols as representing objects and then learn to speak the correct words. In this way, Peter would be taught English, thought Witmer. He went on to predict that the chimpanzees would be taken early in their lives and brought up in homes like human children and, for the purposes of scientific enquiry, taught language skills just as children are.

One of the first experiments was started with Ioni in 1913. Mrs Ladygin-Kohts, living in Moscow, introduced the 1½-year-old chimpanzee into the family. There was no attempt to teach the chimpanzee the Russian language or sign language; the experiment simply recorded the day-by-day events that took place during the next three years in the chimpanzee's development. In 1925 Kohts recorded a further series of observations on the first four years of development of her own son, Roody, and in the early 1930s – quite a while after the chimpanzee work had been completed – she drew up comparisons, which were published in 1935.

Ioni was quite articulate, but not in Russian. The young chimp emitted at least 25 distinct sounds – distinct, that is, to the human ear. They were described by Kohts as emotional responses to particular events in the home. Ioni made no attempt to imitate human speech. Kohts attributed this to the chimpanzee brain being significantly different from that of a child. She did, however, recognize that chimpanzees and human children behave in a similar way during play. They develop comparable emotional expressions, and demonstrate the capability to recognize and identify objects in their immediate environment. Kohts thought that the more important the function for survival, the closer the youngsters appeared to be in their development. When it came to 'higher intellectual' pursuits, the human child was thought to be way ahead.

These conclusions were based, for the most part, on the vocal language capability of Ioni and Roody and the expectation that the young chimpanzee would acquire the use of Russian 'within a matter of months'. This comment on the difference between the basic intellectual wherewithall of chimpanzees and humans was based on the difference between their methods of vocal communication. No account was taken of the possibility that humans and chimps might be of comparable intelligence but that the difference in their methods of communication might mask the fact.

In the meantime, a physician William Furness had tried to bring up other non-human primates in his home. One of his guests was a female orang-utan, obtained from Borneo where Furness had lived in 1909, who, with great effort, was able to make a few sounds that resembled words. Furness manipulated the mouth, lips and tongue of the orang-utan, who could eventually articulate the words 'papa' and 'cup'. The 'kah' sound was made by blocking the animal's nose and pushing a spatula against its tongue. She also added the 'thuh' sound to her repertoire, before dying prematurely just five months after learning her first human sound. Reporting his findings in 1916, Furness described the sounds as hoarse, and thought the orang-utan had considerable difficulty in making them.

In Havana (Cuba) A M Hoyt introduced a gorilla named Toto to his family. In his book *Toto and I: a Gorilla in the Family*, Hoyt wrote about the similarities and differences between Toto's behaviour and that of other home-raised apes, but recorded no instances of vocal communication.

In 1931 a chimpanzee named Gua was born at the Yerkes Anthropoid Experimental Station in Orange Park, Florida but was whisked away, at the tender age of 7½ months, to the home of Winthrop and Luella Kellogg and their 10-month-old son Donald. Gua and Donald were to spend their next nine months living together, the Kelloggs watching their every move.

Gua, the Kelloggs reported, acquired a four-word vocabulary of chimp-type sounds: a general bark, a food bark, a screech, and a soft 'oo-oo' cry. These matched the sounds that R M and A W Yerkes had recorded from captive colonies of chimpanzees. Gua was demonstrating that her vocal sounds were innate emotional responses to particular events and were the same as those of a wild chimpanzee.

The Kelloggs, though, went one stage further than Kohts and tried to teach Gua a few words of English. Gua was intrigued by the facial expressions associated with the word 'papa' but did not try to copy the sound. (Donald, on the other hand, was not doing too well either – he just gurgled and babbled!)

The most famous attempts to teach a chimpanzee to speak was carried out by Keith and Catherine Hayes at the Yerkes Laboratories of Primate Biology in Florida in the mid-1940s. Their star pupil was Viki, a female chimpanzee who joined the Hayes family and lived in their home from just a few days after being born. The chimp was brought up, for the first six years of her life, just like a human baby.

In order to help her learn and speak English words, the Hayes shaped her mouth and lips and rewarded her when she made sounds similar to the ones made by her teachers. In fact, they let her have items of food only after she had made an approximation of the correct sound. They began with a series of simple grunts and progressed to English word sounds. During the course of the experiment, Viki learned to make sounds without the manipulation of her mouth, and was able to conquer just four words. She was able to say the inevitable 'mama' and 'papa' when asked to do so, although the Kelloggs thought she had great difficulty in making the sounds. The other two words – 'cup' and 'up' – were somewhat easier because they resemble sounds that chimpanzees make naturally.

Viki's achievements were not considered great. It became apparent to the Hayes that the chimpanzee's vocal apparatus simply cannot cope with human spoken language. To make two of the sounds, for example, she had first to block her nose (as Furness's orang-utan had done) in order to make the sounds come from her mouth. Humans do this by automatically raising the palate at the back of the mouth to block off the nasal chamber. Chimpanzees cannot do this. Viki also flunked her identification tests, using the wrong words to label objects. The scientific paper describing the Viki experiments was not made public until 15 years later after the work had been completed, and only then as a consequence of publication of the sign-language experiments with Washoe (see page 35).

Then, in the 1970s, Keith Laidler attempted to repeat Furnell's experiments with a young orang-utan at the Flamingo Park Zoo in Yorkshire. The youngster was called Cody (short for Kodjak) and Laidler was to spend a year and a half bringing him up like a human baby and teaching him English. Cody had been rejected by his mother and the zoo authorities had reluctantly separated him from his mother in order that he could get some food.

In his book *The Talking Ape* Laidler describes his first meeting with Cody:

> I don't know what I expected to see. I had never seen a baby orang-utan before, but never in my wildest dreams could I have conjured up the pathetic little bundle before me. There, lying face down on a blanket, was a tiny splodge of auburn hair, not one foot in length.

The study that followed was, at first, a log of the baby orang-utan's early development. During the first hour and a half of observations, Laidler was to discover a slight drawback to working with young apes: like human babies, they sleep for most of the time, and when they awake they are interested only in food.

Cody's rather spartan zoo surroundings were converted into the equivalent of a children's playroom. He wore nappies, slept with a teddy bear, and played children's games. His speech lessons followed the same pattern as those pioneered by the Hayes with Viki, using reward and mouth shaping as a means of developing any sounds he might make.

Laidler discovered that the best way to make Cody emit a sound – any sound – was to move away. Cody was reluctant to let his 'foster mother' go far. If Laidler moved, particularly out of eyeshot, Cody shouted, and for his shout he got a reward.

Compared to a chimpanzee or a gorilla, the orang-utan was a slow developer; though he was faster than a child. This middle-man position of the baby orang-utan interested Laidler. He thought it might indicate a greater intelligence than that possessed by the other apes. Cody's sound-making ability, however, was not at all promising at first. In desperation, Laidler adopted Furness's technique of trying to get the youngster to breathe through his mouth instead of his nose. Cody had other plans: he would refuse to breathe through his mouth, go bright red in the face, and squirm out of Laidler's grip.

About this time all speech training came to a halt. Cody was taken ill. After two weeks of diarrhoea, soiled nappies, and the fear – because gastric enteritis in young apes can be very serious – that Cody might die, the youngster began to get better, and Laidler took stock of his experiment. So far, despite three months of training, Cody had said nothing. Laidler took his problem to the speech therapy department of Newcastle University and asked for advice. Margaret Edwards was in charge, and she came up with a speech teaching regime that she thought might work with an orang-utan. It was designed for use with autistic children and was based on the reinforcement methods devised by the distinguished psychologist B F Skinner. Skinner and his followers believe that all behaviour is controlled by expectation of reward or punishment, pleasure or pain, positive or negative reinforcement, carrot or stick. If an animal does anything that remotely resembles the required action, it is rewarded. Gradually, its behaviour can be moulded, each step taking it nearer the intended goal.

Of the several options open to him, Laidler chose a method of speech therapy devised by Frank Hewett, who had had great successs in teaching autistic children to talk. The method is pretty tough. It requires a wooden booth with two cubicles. The pupil sits in one cubicle that is not lit and which can be shut off from the teacher's cubicle by a hatch, much like a serving hatch. The pupil is positively reinforced with the opening of the hatch (food being presented) and contact with the teacher. A tantrum or a refusal to cooperate is met with a closed door, no food, no light and no contact. It worked well with autistic children. Indeed, Laidler thought it almost miraculous the way that mute children could be taught to communicate. But would it work with an orang-utan?

Cody would have nothing to do with the double-cubicle. He went berserk every time he was separated from his teacher and 'mother'. Laidler had to settle for a modified version that consisted of a six-feet-tall square box in which they both sat, Cody on Laidler's lap. The advantage was that the box removed all outside distractions and Cody could concentrate on Laidler's hands and face.

At first, to introduce him to the teaching regime, Cody was not taught words but simple actions. He was taught, reluctantly at first, to put his hand on his head. The reward was a Cadbury's Chocolate Button (young orang-utans seem to have a passion for chocolate). It did not, however, always work, as a bruised and bloodied Laidler discovered when Cody landed him a left-hook on the nose and knocked him clean out of the booth – right in front of an astonished elderly lady who was peering into the previously empty cage.

In order to stop the training sessions from getting out of hand, Laidler had to change tactics. He introduced 'punishment periods', known in Skinner-speak as 'time out from positive reinforcement'. The miscreant is denied his or her favourite activity. The time imposed gets longer the more the pupil misbehaves. Cody was pinned to the floor, first for 15 seconds and then gradually longer, in order that he could not play. He quickly knuckled under. After 200 hours of tuition and 53 packets of chocolate buttons, Cody complied with the hand-on-head movement. If he could imitate movements, thought Laidler, it was quite possible that he could imitate sounds.

A few days after this breakthrough, they were approaching the booth when Cody spontaneously put his hand on his head. He was using a sign to indicate he wanted food. At this stage, Laidler thought he could have easily taught Cody sign language, but he had made the decision to teach speech and so, after withdrawing the hand-on-head movement as a part of the training, started along the road to speech training.

First Cody had to learn to breathe through his mouth. This turned out to be no easy task. Laidler had to put his hand over Cody's nostrils in order to force him to use his mouth. Cody was obstinate, holding his breath for up to a minute on one occasion, before eventually taking his first gasp through his mouth. Gradually, the time between gasps was reduced and Cody breathed in the correct way on demand. But Cody was to surprise Laidler again. In addition to the gasp, he made a 'gruh' sound. He was rewarded each time with a chocolate button, and the 'gruh' was changed to a 'kuh'. By seven months of age, he could 'kuh' within a few seconds of his nose being closed. Not long after, Cody learned to block his own nose and say 'kuh'; even more significant, he could sometimes make the 'kuh' sound without closing his nose – something neither Viki nor the orang-utan of Dr Furness could do.

Again, Cody anticipated teaching sessions and made the 'kuh' sounds before they had reached the training box. Cody had learned to associate the sound 'kuh' with the expectation of chocolate buttons. He also used the 'kuh' sound while playing. He seemed to be practising it.

Then Cody was encouraged to move outside the box and respond to teaching sessions in the middle of the room. He was encouraged to use the 'kuh' sound for another action – to open the connecting door between his living compartment and the teaching area. He began to associate the sound with things he wanted – food, tickling, door-opening, and hugging. It was time for Cody to learn his second sound.

Laidler chose 'puh'. Unlike 'kuh', which is produced from the back of the throat, 'puh' is produced by closing the lips as air is expelled. It is, therefore, a very different sound for the young orang-utan to learn. Laidler moulded the youngster's mouth, and after learning to imitate a pout, Cody produced a more-than-adequate 'raspberry' or 'Bronx cheer'. Unfortunately for Laidler, Cody's mouth was full of food at the time and Laidler was presented, rather unceremoniously, with a face-full of sticky oat flakes. Laidler continued to reinforce the raspberry-like 'puh', but had to 'undergo the indignity of being sprayed with Farex and honey twenty times a session'.

Not long after, Cody became confused and the 'puhs' and 'kuhs' got muddled up. Laidler was understandably depressed, but he persevered and sorted it out by starting the 'puh' training all over again. Laidler interpreted the mix up as a challenge to the original 'kuh' sound. Before 'puh' came along, Cody was able to get everything he wanted with 'kuh'. 'Puh', thought Laidler, caused some conflict in the youngster's mind. Human children appear to do the same with each new word they learn. At first, a word like 'miaow', used originally for cat, can be applied to all sorts of four-legged, furry animals including bears. But, as a child gets older, 'miaow' is confined to a domestic cat, before being replaced altogether with the proper word, 'cat'. Cody was having the same problem with his universal 'kuh' sound. Eventually, after further training, he was to learn that 'puh' and 'kuh' were to be used for different things – 'kuh' for drinks and 'puh' for food.

Cody became a television star. In exchange for the right to continue the experiment at Durham University, the zoo authorities gleaned maximum publicity from their 'talking ape'. Meanwhile, he learned his third sound, 'fuh'. It was to replace the 'puh' sound to represent food. The 'puh' sound was switched to be associated with being picked up and hugged. Cody was now more than a year old.

When the summer season had finished, pupil, teacher and

newly wedded wife moved to Sunderland. Cody was ready for his fourth sound, 'thuh'. This required the orang-utan to put his tongue between his teeth and expel air. Laidler was running out of objects for the word to represent as Cody already had words for his favourite activities. Then, during a belly-brushing session, it became clear in what context the new word could be used: 'thuh' would signify brushing. He mastered 'thuh' in just three days, in contrast to Viki the chimpanzee, who had taken months rather than days to learn a new sound.

By this time, Cody was using the four words not in a teaching box with a single teacher but in a variety of situations and with different people. Each word was used correctly for 70 per cent of the time. He had demonstrated that he understood the concepts of food and drink and was able to associate these with articulated words, something no other non-human primate had achieved.

He also sometimes used the words in a novel way. At first, if he bit anybody, he received a slap and, suitably chastened, just left it at that. But later in the experiment, a slap was followed by Cody jumping onto Laidler's knee and uttering 'thuh', the sound for 'brush me'. It was, Laidler thought, Cody's way of saying sorry.

On another occasion he appeared to invent a word. Laidler's mother had been 'orang-sitting', when Cody pulled her into the kitchen in order that she might find him some food. On the way, he kept making a 'sheesh' sound. Did this, wondered Laidler, mean 'follow me'?

Cody also used his words in intriguing ways. Once, he was playing on the floor when Laidler brought him some cereal and honey. Cody said 'puh', meaning 'pick me up'. Laidler was expecting Cody to say the word for food and asked Cody what he had said. Cody started to climb his teacher's leg and continued to say 'puh'. On reaching the top half of Laidler's body, Cody sat on the right hip, looked straight at the bowl, and only then said 'fuh', meaning food.

There was, however, one peculiarity of Cody's vocalizations that perplexed Laidler. Cody did not distinguish between animate and inanimate objects. He spoke to doors, tables, the window and even a piece of wood.

There was one more event that was to have a galvanizing effect on what Laidler should do next. The young Cody was by now a powerful animal and at some stage, it was thought, he should return to the zoo. Laidler took him back to his old quarters occasionally just so that he would remember them. On one visit, Cody was taken to the front of his mother's cage. The older orang-utan put her arm forward in greeting. Cody shrank back in terror: he had become humanized to the point of being frightened of his own mother. Laidler decided that the quicker he was introduced to other apes the better. The experiment was brought to an abrupt end. On separation day, Laidler waited until Cody was playing with Freddy the Teddy and quietly slipped out.

Talking Hands

Long before Charles Darwin's revelation that our ancestors were once ape-like creatures and that we are closely related to the surviving great apes, several writers had observed the disconcerting similarities between us and them and had speculated on whether we might be able to talk together. The emphasis, as usual, was to teach great apes to speak to us in our language rather than we speaking to them in theirs.

In the daily meanderings of *The Diary of Samuel Pepys*, the great diarist reflected on this very subject on 24 August 1661; the diary entry for that day included the following:

> At the office in the morning and did business. By and by we are called to Sir W Batten's to see the strange creature that Captain Holmes hath brought with him from Guinea; it is a great baboon, but so much like a man in most things, that (although they say there is a species of them) yet I cannot believe but that it is a monster got of a man and a she-baboon. I do believe it already understands much

> English; and I am of the mind that it might be taught to speak or make signs.

Robert Holmes, the captain of a ship, had sailed to West Africa and it is thought that the creature he brought back was a chimpanzee or a gorilla. Tales of interbreeding between people and other animals were common in those times, so it is no wonder that the diarist suspected the worst. But, whatever the identity, it is significant that Pepys thought we might be able to communicate with it, perhaps by using sign language.

In the mid-eighteenth century Julien Offray de La Mettrie, working in France, was also struck by the similarity of the great apes and some of the monkeys to ourselves and suggested that we find a means by which we might converse. He was certain, even at that time, that apes and monkeys did not have the capability to make human sounds and form human words, and therefore he, too, suggested the use of some kind of sign language. Apes and monkeys, he thought, have similar hands to those of humans, and the fingers move in roughly the same way. Why not make use of ape and human dexterity to create a new language that both might use? After all, sign languages of one sort or another have been used by people for thousands of years.

It is likely a gestural language was used by early man, particularly during a hunt when sounds would have alerted the prey. Certainly, more recent hunters, such as the various tribes of the Plains Indians of North America, used a well-documented sign language; with each tribe speaking one of six basic families of languages, the only way different tribes could talk to each other was by a sign language that was common to all. Likewise, the Bushmen of the Kalahari and the Aborigines of Australia have gestures by which individuals in the tribe can communicate at times when making a sound is inappropriate or inconvenient. The Bushmen really did 'speak' with fork tongue: their hunting sign for springhaare is the familiar two-finger victory salute.

And, in western society there is the frantic tick-tack language used by bookmakers at horse-races, and the signals used by referees and umpires in sports such as basketball and cricket. Sign language, whether the upraised middle-finger exchanged by irate Italian car-drivers or the clenched fist raised above a political activist's head, is still a powerful way of putting over your message.

But in those early research days, not long before La Mettrie made his proposal, J C Amman, another early researcher, had gone some way to creating a form of communication which could be used with people who were either deaf or mute. He drew up plans for teaching signs, finger spelling and lip reading. In 1620 a Spaniard, Juan Pablo Bonet, described a deaf and dumb sign language; Bonet wrote the first book on the subject, based on his own work and that of his predecessor, Pedro Ponce de León. In Britain in 1635 a Mrs Babington of Burntwood, Essex, had also invented a sign language to converse with her deaf husband. La Mettrie must have been acquainted with Amman's work (and maybe even the sign languages developed by Bonet and Mrs Babington) and in his book *L'Homme machine*, which deals with the extent to which animals can learn, he went as far as to write:

> Why should the education of monkeys be impossible? Why might not the monkey, by dint of great pains, at last imitate after the manner of deaf mutes, the motions necessary for pronunciation?

Again, the emphasis was to teach animals to talk to us in a way which we had invented; there was little interest in studying the way they communicated and learning their grunts and gestures.

Nobody seems to have followed up La Mettrie's early proposals for teaching an ape sign language, although Charles-Michel, Abbé de L'Epée, considered it an appropriate way to converse with the deaf or hard of hearing. He founded the first school for the deaf in Paris in 1760, and introduced a sign and finger language. This was further explored and developed by the Abbé Sicard at the end of the eighteenth century, and eventually the manual or silent method of teaching the deaf was created. As for communication by sign language between ape and man, the matter rested there until 200 years had passed and the distinguished primatologist R M Yerkes wrote in his *Chimpanzee Intelligence and Its Vocal Expressions*:

> I am inclined to conclude from the various evidences that the great apes have plenty to talk about, but no gift for the use of sounds to represent individual, as contrasted to racial, feelings or ideas. Perhaps they can be taught to use their fingers, somewhat as does the deaf and dumb person, and helped to acquire a simple, nonvocal sign language.

And, again, nobody took up the challenge until 40 years later, when the Gardners – Allen and Beatrice, professor and research associate respectively in the psychology department at the University of Nevada, at Reno – began a series of experiments with a chimpanzee who was to become one of the most famous of her kind. Her name was Washoe.

Washoe was caught in the wild in the mid-1960s (a practice which nowadays is frowned upon, if not condemned) and was 8 to 14 months old, a period in a chimp's life when it is still dependent upon its parents – which probably meant that Washoe's parents had been killed in order that the baby could be taken into captivity. The Gardners obtained her from an American animal trader. This orphaned baby eventually achieved greatness, and became the first non-human animal with which people could talk.

With the failure of Peter, Ioni, Gua and Viki to talk to their keepers using a vocal language, and the realization that chimpanzees do not have the vocal apparatus to form the complex sounds of human words, the Gardners turned to gestures as a way to communicate with Washoe. Many field researchers – such as Jane Goodall at Gombe (Tanzania) and

Adrian Kortlaandt in the Congo and Guinea, Vernon and Frances Reynolds in Uganda, Junichero Itani in Tanzania – and others, such as Jan and Anton van Hoof at the Burgers' Zoo, Arnhem, studying captive animals, had noticed that chimpanzees use their hands in greeting or aggressive displays. Here, the Gardners thought, was a clue to success, and Washoe became the first chimp to be taught a human gestural language; in her case, American Sign Language (Ameslan or ASL).

First, the Gardners and everybody else involved in the experiment had to learn ASL, the idea being that in the presence of Washoe this would be the only form of communication used: nobody, working or playing with Washoe, would talk normally. Sounds were allowed only as spontaneous bursts of surprise or enjoyment. It would have been undesirable for Washoe to associate vocal communication with 'adults', and sign language only with 'babies'; so everybody used ASL.

Washoe lived in a specially-built 2.5 x 7.3m (8 x 24ft) caravan in the Gardners' 465sq m (5,000sq ft) backyard. During all her waking hours she had at least one, and usually many more, human companions. In the yard she climbed trees and played on children's swings and climbing-frames. Sometimes she was allowed to visit the Gardners' home, and occasionally she was taken on excursions in the car. Her confinement (which if she had been kept in a cage would have resembled the barren social conditions that were once experienced by patients kept in isolation in old mental hospitals) was kept to a minimum. She always had playmates and protectors. Teaching sessions became a part of the daily routine. It was important that Washoe lived in a rich and stimulating environment. She was never bored.

The experiment started in June 1966 and by October 1970, when the initial project ended, she had acquired a vocabulary of over 130 signs. That may not sound much in human terms, but Washoe was able to put the signs together with considerable inventiveness – they were syntactically interesting and contextually correct. She could also use her signs to describe situations with which she had had no previous experience.

Washoe acquired her signs in a number of ways. There were 'shaping' or 'conditioning' experiments in which she was rewarded each time she completed a task or learned a new sign. Her favourite reward was to be tickled and she would often make the sign for 'tickle me'.

For some signs, Washoe's hands were initially guided into the correct positions and movements – a process known as moulding. This proved to be the fastest learning method, and indeed her 'tickle me' sign – which was her sixth sign – was taught this way. It consisted of the index finger of one hand brushed across the back of the other hand, which was open and palm downwards. During the learning process, the Gardners reinforced the sign by making the gesture themselves, guiding the chimp's hands to do the same, and then – to Washoe's sheer delight – tickling her.

There was also the simple expedient of creating a sign for every event or task that was performed during the day. Washoe remembered some of them, simply by seeing them regularly. She slowly came to associate specific signs with certain objects or activities. For example, she hated cleaning her teeth. But after every meal she was encouraged to do so. The sign was simply the movement of the index finger across the teeth. Imagine, then, the surprise when Washoe came to visit the Gardners, went to the bathroom, climbed onto the wash basin and, sitting beside a rack filled with the Gardner family toothbrushes, made the sign for 'teeth cleaning'. Washoe certainly had no intention of cleaning her teeth but had simply wanted, much like a stranger at a cocktail party seizing on a familiar object or event, to strike up a conversation. This happened about 10 months into the experiment.

Just as human babies 'babble' vocally, so too did Washoe. She 'babbled' the sign language equivalent of 'goo-goo' with her hands. She 'babbled' very little at first but, after she had begun to acquire a few 'words', she increased her 'babbling'. She began to experiment with signs and was encouraged by her trainers until her 'babble' turned into the proper sign. After two years, when her vocabulary had become quite extensive, 'babbling' ceased altogether.

Imitation was used to correct and speed up signing and to improve on Washoe's 'diction'. If the chimp made a poor sign, the trainer exaggerated the signal to demonstrate to Washoe what she should do. Unfortunately, Washoe was sometimes on a short fuse and threw an enormous tantrum when 'told off' or corrected. Likewise, she became angry when she could not find the right sign for a particular event and swung her arms round in frustration. She also invented her own signs – a beckoning gesture for 'come-gimme' and a violent shaking of the open hand for 'hurry'. She even invented the sign for 'bib' by drawing a square on her chest with her finger.

Washoe was a slow starter but eventually was considered to be a fast learner. In the first seven months she learned just four signs; in the second seven months there were nine new signs; in the third there were a further 21. And the more she learned, the more she was able to differentiate.

At first, when Washoe was attracted by the smell of something, she would make the sign for 'flower'. Later in the experiment, she distinguished cooking smells from the scent of flowers with two quite different signs. In addition, her sign for 'flower' became generalized and was used for all flowers, indoors or out, real or painted. She had clearly recognized things for what they are, rather than associating them only with the specific objects employed in the experiments.

Washoe was able eventually to combine signs into simple phrases. At first, for example, she would bang on a door to gain entry. Then, she was taught the sign for 'open' – two open palms placed against the object to be opened and then moving them up and apart. She soon learned to use this not only with doors but also books, boxes and even briefcases.

Finally, she was able to combine the sign for 'open' with something else. She would approach the refrigerator, for example, and sign 'open food drink'; and when an alarm-clock sounded at mealtimes she would signal 'listen eat'. Washoe also told of her future intentions with 'go-in', 'go-out', and 'in down bed', and she distinguished between various members of the team with 'Roger tickle' and so on. And, on learning the word for 'hat', she seized on a brightly coloured plastic bag, placed it on her head, and ran about signing 'hat, hat'.

And even more interestingly, Washoe differentiated between 'I-me' and 'you'. The animal was conscious of itself and said so, and in doing so contributed, albeit in a simple way, to the philosophical debate taxing the great minds in the human world on the concept of 'self'.

As in all experiments of this sort, the subject had to be tested, simply to prove or disprove that she was actually acquiring and was able to use ASL. The tests were double-blind trials. The trainers were not allowed to see the tests for fear that they might unconsciously 'cue' the animal to make the correct response – the 'Clever Hans syndrome'.

Washoe did tolerably well with pictures, models in a box and colour slides. In a test with 35mm colour slides, Washoe was asked to identify objects and was able to label 53 out of 99 correctly. That does not sound very impressive; but what interested the Gardners was not the signs Washoe got right but the ones she got wrong, for in these cases the signs she used, although actually incorrect, were conceptually significant. Washoe signed 'dog' when the picture was of a cat, 'brush' for a comb, and 'food' for meat.

There were also interesting results when the same object was presented as a toy or the real version. Washoe's response to a model toy dog was to sign 'baby'. She also signed 'baby' to a picture of a toy dog. But to the picture of a real dog, she signed 'dog'.

As a whole, the observations were intriguing, sometimes surprising, and often amusing. But how significant were they in the language stakes? Was Washoe using language? The Gardners were understandably cautious in the way they analysed their data. It was clear that Washoe had preferences for the order of words; 'you me' took precedence over 'me you'. And, in three-word combinations, 'you-me-action' was first preferred, although later in the study 'you-action-me' was the more usual order of signs. Was this evidence of syntax or simply an imitation of a preferred order of words introduced, perhaps, by a member of the team? Whatever the truth, some researchers believe there is little difference between the way Washoe selected her order of words and the way human children learn syntax.

Indeed, in one exchange recorded by the Gardners the similarity between the chimp talking to its trainer and a mother talking to a child is marked. The conversation in ASL went like this:

Washoe: Gimme.
Trainer: What this? [i.e. what do you want?]
Washoe: Food. Gimme.
Trainer: Ask politely.
Washoe: Please.

The sign 'gimme' was one of Washoe's key verbs. Pioneering psycholinguist Roger Brown gave these frequently used child verbs the name 'pivot' words, and they are often used with nouns which he calls 'open' words. A child's 'pivot' word might be 'go', and it might be used with an 'open' word, to give 'go shop' or 'go toilet'. Washoe frequently used 'gimme food'.

In fact, Washoe was so skilled at using ASL, she became quite devious and tried to cheat her trainers. Using the 'open' sign beside a cupboard, she waited patiently for the cupboard to be opened and then grabbed at some object with which she was normally forbidden to play. A few moments later, she signed 'open' again, but was refused. So Washoe went to her potty and produced, after a great deal of effort, a small amount of urine. The trainer, then, had to open the toilet door to empty the potty and, just as the door was opened, Washoe ran in and grabbed even more forbidden objects.

But perhaps her most entertaining sign sequence came later in the experiments, when she was introduced to a number of different monkeys and apes at a laboratory. When she arrived she had a violent disagreement with a macaque and was whisked away to look at a couple of siamang gibbons and taught the sign for 'monkey'. She was quick to learn and, quite soon after, signed 'monkey' to some squirrel monkeys. Then Washoe was presented once more with the quarrelsome macaque to which she signed 'dirty monkey'. Previously, the sign 'dirty' was an 'open' word which represented soiled articles or faeces. Washoe had changed it into an adjective and invented her first insult.

Washoe's initial experiments came to an end in October 1970, and she was removed from her cosy caravan in Nevada and institutionalized in Oklahoma. She accompanied Roger Fouts, one of the Gardners' students, to the Institute for Primate Studies in Norman, where she joined a whole group of sign-talking chimpanzees, some in a main colony and others brought up in private homes. Here the chimps, like children in a school, were being put through the rigours of teaching and testing. For them it was not the three 'Rs' that were important as their ability to pick up ASL.

Not unexpectedly, chimpanzees, like children, learn at different rates and are good at different things. One young female with a depressingly low score, sure to give a human parent palpitations, would do well only if continually praised. In the double-blind trials, where her trainers were unable to pat her on the back when she got things right because they couldn't see the experiment, she was relegated to class dunce. There was, however, one star at Norman. She was called Lucy and, like Washoe, she was to make a considerable contribution to man-chimpanzee communication.

Lucy was also raised in isolation from other chimpanzees,

having had only human companions from the age of two days. Her big day came when she was seven years old and, in the company of Roger Fouts and her foster 'parents', psychoanalyst Maurice Temerlin and his wife Jane, she was about to show whether a new sign would be used for a specific object only, or whether it could be used to represent similar or related objects.

Lucy had already had a 75-sign vocabulary and had learned the generic food signs 'food', 'fruit', and 'drink', and had acquired the specific signs for 'candy' (sweets) and 'banana'. The new sign was to be 'berry' and Lucy was first taught to associate the sign for 'berry' with a cherry.

She was presented with food items mixed up with non-food items and was asked to identify them. She had a sign for all the items shown to her. During the first few days of the experiment she was told the cherry was represented by 'berry', but after a while blueberries were substituted and she was instructed to sign 'berry' to the blueberries. For a couple of days after that, Lucy continued to sign 'berry' to blueberries but then reverted to the 'berry' sign only to cherries. The experiment, the researchers felt, demonstrated that chimpanzees prefer to use their specific words in a specific sense and they will not be persuaded otherwise.

There was, however, more to come. Lucy came up with some extraordinary sign-language descriptions of the fruits and vegetables with which she was presented. Watermelons were good news – they were labelled 'candy drink' or, even more accurately, 'drink fruit', despite the fact that the researchers used quite different sign words, such as 'water' and 'melon' that Lucy did not know. Citrus fruits, with their very strong aromas, became 'smell fruit'.

The funniest example, though, was her encounter with the radish. At first, she examined the little vegetables but didn't eat them and signed quite rightly 'fruit food'. On the fourth day, she took the plunge and bit into the radish, immediately spat it out, and signed indignantly 'cry hurt food'! These novel combinations of sign language words showed that Lucy could use her existing vocabulary to create new concepts to encapsulate her understanding of the objects with which she was presented. She was, in effect, demonstrating an ability to think in the abstract.

Booee and Bruno, two young male chimpanzees with a sign language ability of 36 signs each, were the next characters to take the stage. Until their experiment, all the communication had been between humans and chimps. But, for a language to be a true language, it must be used by animals of the same species. At first, Booee and Bruno, however, preferred their own natural communication over ASL, although they regularly used the sign for 'hug me' when comfort was sought. Booee once asked Bruno to 'tickle Booee' but was firmly rebuffed with the sign sequence 'Booee me food'. Bruno was far too interested in the apple he was eating than to be fussed with tickling Booee.

Several sign-language experiments with chimpanzees followed the Washoe project. The Gardners introduced four young chimpanzees to semi-deaf trainers who were more adept at using ASL than people with normal hearing. There was some concern that the inability of chimpanzees to progress further was due to the widely differing abilities of the human trainers to communicate in ASL – there were considerable inconsistencies in the signing procedures between trainers, which could have confused the pupils. By having trainers who had been brought up themselves with ASL, there would be more continuity between training sessions.

And Washoe herself was responsible for training new young chimps. She went with Roger and Debbie Fouts to the primate facility at Central Washington State University at Ellensburg, where she taught ASL to a foster chimp called Loulis. The keepers were careful to use only five signs while in Loulis's presence, so most of the signs the youngster acquired were learned from Washoe. Washoe even copied the Gardners' original teaching methods. She continually made the sign for an object, like her human teachers had done. She even took the youngster's hands in hers and 'moulded' them into the right sign shape.

Previously, some time before, Washoe had lost her own first baby. She sat near the inactive corpse continually signing 'baby, baby'. Was it simply a sign of recognition or a more poignant sign of grief? Later, she gave birth to a son, Sequoyah, and started to sign to it.

Koko and the Cat

Until recently, the gorilla had received a bad press, usually depicted as a fearsomely powerful and aggressive beast. From the time of its discovery by the American medical missionary Thomas Savage, as recently as 1847, the gorilla had been misrepresented. The skull that Savage examined confirmed ancient reports from the fifth century, when the Carthaginians sailed along the west coast of Africa and killed the 'hairy people – the gorillae'. The Franco-American adventurer Paul du Chaillu travelled in Savage's footsteps in 1856 and followed tradition. Confronted with an enormous male, he did what seems to come naturally to our species: he shot it. Today, through the field work of George Schaller, Dian Fossey, and their students, we have come to know a lot about the social life of what, in fact, are gentle vegetarians. And a project that began in 1972 in San Francisco placed these supposedly savage beasts in a wholly different light.

The experiment followed in the tradition of the Gardners and Washoe, but the pupil was a captive-born lowland gorilla called (as a result of a local competition) Hanabi-Ko, meaning in Japanese 'Fireworks Child'. They called her Koko for short. Her name was chosen because she was born on American Independence day. The place was San Francisco Zoo, and the date 4 July 1971.

Koko was the daughter of Jackie and Bwana, but Jackie was a mother with little milk and so when dysentery swept through the gorilla enclosures, the baby had little defence and came close to death. She was six months old, yet weighed only 2.2kg (4lb 14oz), the normal weight for gorillas at birth, not at the end of their first half-year. She was suffering from malnutrition, had diarrhoea, was hairless and dehydrated. She was rushed to the Animal Care Facility of the University of California Medical Center, and was shortly returned to the zoo to be looked after by the families of the zoo director and of the manager of the Children's Zoo. Koko recovered well and went on permanent exhibition in the Children's Zoo.

In the meantime, watching the little gorilla from the sidelines, was a young researcher by the name of Francine (Penny) Patterson. Attending a lecture at Stanford University given by the Gardners, she became so intrigued by the research they had being carrying out during the previous five years with Washoe that she determined to devote her work to 'the language abilities of animals'.

At that time Patterson was working with Stanford neuropsychologist Karl Pribram, and he was discussing with the zoo authorities the prospect of setting up a special enclosure where a device could be built by which gorillas could be encouraged to talk via a system of keys. On a visit to the zoo, Patterson spotted Koko and thought that she could attempt an ASL experiment with a gorilla just as the Gardners had done with a chimpanzee. Patterson's first proposal was turned down; the zoo was understandably more interested in breeding endangered species than carrying out any experiments that would involve separating the baby from its mother. Undeterred, Patterson waited for another gorilla, another time, and learned ASL.

Then Koko became ill and had to be separated from her mother anyway. At last an opportunity presented itself and Patterson was quick to suggest it to the Zoo Director. He agreed; Dr Pribram agreed; and the project, with no funding, few resources, and no schedule, was underway.

Penny first met Koko, quite literally face-to-face, on 12 July 1972. The little gorilla pushed her face into Penny's and sniffed around. Penny put Koko down onto the floor and made the sign for 'hello', much like a military salute. Koko responded by putting her hand on her head and did what any mischievous gorilla might do in the circumstance – she pulled Penny's hair. Penny was to learn that being a dedicated scientist interested in teaching another primate sign language *and* being mother to a one-year-old, 9kg (20lb) gorilla, was no easy task.

Koko's first signs were to be 'drink', 'food' and 'more'. Penny did not fill Koko's baby bottle to the top, but just half-full, hoping anticipation of the other half would be sufficient stimulus to use the signs. Koko had a different idea – she just ignored the signs, grabbed the bottle and sucked on it until not a drop was left.

But, within two weeks of the start of the experiment, Koko began to make approximations of the 'food', 'drink', and 'more' signs. Penny was not impressed at first, regarding them as random gestures without context. Then, on 9 August, Koko consistently made the sign for 'food' in response to some pieces of fruit. Penny's enthusiasm for Koko's success meant that the poor gorilla was stuffed full of food and, by the end of the day, had absolutely no interest in the smallest crumb.

The young gorilla was quick to learn new signs and began two-word combinations by 14 August with 'gimme food'. A few days later, she put two and two together by calling her babyfood drink – a mixture of cereal and milk – 'food drink'. For the next few weeks she continued to use her small vocabulary in many different combinations. Then, three months into the experiment, Koko made another giant leap: she asked questions.

When using ASL, deaf people ask questions either with their eyes or by moving their hands in a certain way, and this is just what Koko did. Her first attempt was in the form of a request. Penny breathed onto the cold window of the enclosure and showed Koko how to draw in the film of condensation. Koko asked her to do it again. She pointed to Penny's lips and looked her straight in the eye. Later Koko took to cocking her head to one side in order to change a signed statement into a question. On one occasion, a woodpecker was drumming outside the caravan. Koko's teacher signed 'Koko, listen bird'. Koko held the sign for 'bird', looked into her teacher's eyes and raised her eyebrows. She had turned 'bird' into 'bird?' and, having established that she had understood, Koko finally gave the woodpecker its new gorilla-style name and signed 'listen bird'.

The other intriguing thing about Koko's progress was her inventions and the way in which she deviated from the signs she had been taught. Penny and her research assistants quickly realized they had two experiments running simultaneously – first was the attempt to teach ASL and discover whether Koko could generate words and understand them; second was the use made by Koko of the things she was being taught. In order to make the data scientifically valid, Penny was at great pains to build up a set of statistics from the hours of videotaping, logging and testing, so that an independent researcher would be able to see immediately the context and conditions in which Koko used a word. It meant hours of laborious transcribing.

By October Koko was coming on apace, but there were some serious limitations on what she was able to do. Just as chimpanzees could not talk vocally in English because the vocal apparatus in the gorilla throat is the wrong shape, so too there is problem with gorillas' hands. Some of the ASL signs are impossible for a gorilla to perform. The thumb, for example, is much smaller than that of a human and it is positioned much further down the side of the hand. Gorilla variants of these human ASL signs had to be found.

Koko, like Washoe before her, had trouble making signs that involved placing the hands away from the body. The sign for 'milk', for example, is simply a fist held in front of the body and squeezed, as if milking a cow. Koko made the sign against her chest. And learning the signs for objects she did not care for tended to be hard going. The signs for 'lollipops', 'swing' and 'berry' were all easy, but 'egg', which Koko disliked intensely, was not an easy one to teach her.

She also teased her teachers. Once, when asked to sign 'drink', Koko used every appropriate word she knew, from 'sip' to 'thirsty sip' to 'apple sip', but she stubbornly refused to sign 'drink'. After a while, the exhausted and desperate teacher pleaded with Koko and said 'Please, please sign "drink" for me'; at which Koko sat back, grinned broadly, and signed a perfect 'drink' sign – but to her ear. And there was the occasion she was asked to place a toy 'under' a bag. Koko, quite poker-faced, picked up the toy and raised it to the ceiling. It demonstrated, thought Penny, that Koko was capable of 'verbal playfulness'. The exaggerated way she performed the tasks seemed to show she knew very clearly what she was doing. They were not mistakes.

Indeed, later in the experiment, Koko discovered the art of insult with signs for 'dirty', 'bird', and 'nut'. One of the deaf teachers who had been working on the project had had an argument with Koko. The incident started when Koko had been shown a picture of herself on a poster.

Teacher: What's this?
Koko: Gorilla
Teacher: Who gorilla?
Koko: Bird
Teacher: You bird?
Koko: You
Teacher: Not me, you are bird
Koko: Me gorilla
Teacher: Who bird?
Koko: You nut
Teacher: Why me nut?
Koko: Nut, nut
Teacher: You nut, not me
Koko: Damn me good

At that point, Koko walked away signing 'bad'. Penny was intrigued by the way she could insult, joke and even lie.

Koko was, thought Penny, exploiting language just as we do – an ability that should have been beyond her, according to the experts.

It was generally believed that lowland gorillas in their natural environment in the African rain-forests were no more than the primate equivalent of cows. Certainly they seem to have an undemanding life, languidly munching their way through wild celery, nettles or bamboo shoots, with little need to consider the future in the way a meat-eating hunter might have to do. Cooperative hunting needs some degree of planning and even discussion.

Gorillas, though, have something else – they have long-lived social groups. The ability to 'think' ahead could be important in amalgamating the conflicting pressures of competition and cooperation that exist in such a group. Indeed, they have their own 'language' of gestures and grunts which are clearly important in a close social society. It could be this social need that allows the gorilla, and the chimpanzee, to do things in a 'language-rich' captive environment that it would never need to do in the wild – such as speaking with a sign language.

Koko's acquisition of new signs grew rapidly – at about the same rate, in fact, as Washoe's. By the end of 18 months she had learned 22 signs to Washoe's 21.

In June 1973 Koko moved house. She was getting too big for the children's zoo. A woman had banged on the glass, as people tend to do in zoos and aquaria despite numerous notices asking not to, and Koko had banged back – but a little too hard, and had broken the glass. Her new home was a 50 by 10ft caravan next to the zoo's gorilla grotto. It had sleeping and living accommodation for both keeper and gorilla. During her first few nights, just like a human child who has moved house, Koko had nightmares and had to be comforted. She also had to be pot-trained (the trailer was carpeted throughout) and, like her human equivalent, discovered that her keeper's interest in toilet training could be used to her advantage. Apparent lapses were interpreted as attention-seeking.

Lessons continued. After the initial learning period, Koko became bored with learning and signing the same word day after day. This was a testing procedure copied from the Gardners. Washoe had been asked to perform the same word 14 days running in order to prove that she knew it. Koko, like Washoe, reached a phase in her education where that system became inappropriate. Psychologists call it 'learning to learn': the time a child learns something and then, having mastered it, wants to get on with something else. What can one make of the fact that Koko could sign 'cucumber' for 13 days on the trot and then flunked on day 14? There were getting to be grey areas in the research. Penny got round this by requiring Koko to sign successfully before two independent witnesses on at least half the days in the month.

Koko was also getting a bit cocky. Koko was once playing by herself with some white towels and her teacher noticed she was signing 'red'. The teacher said 'You know better, Koko. What colour is it?' Koko continued to sign 'red', each time exaggerating the gesture for the full effect. Eventually, with a big grin, she picked up a small speck of red material that had been stuck on one of the towels and, thrusting it under the teacher's nose, signed 'red'.

When Project Koko was in its second year, the return of Koko to the gorilla grotto for breeding purposes was uppermost in the minds of the zoo's administrators. Penny was of two minds. At first she had considered working for about four years with Koko and then returning her to the zoo's other gorillas before she got out of hand. It was common belief, at the time, that both chimpanzees and gorillas become unmanageable after a while. But during the early months of Project Koko, Penny began to wonder whether it was really necessary for Koko to have a baby in the zoo. They could get her a companion and the experiment could continue. Penny had also seen several people romping with full-grown gorillas and felt that, if they could do it, so could she. She was also getting emotionally involved – Koko was her adopted baby, after all.

They played games. Koko hid under a fold-up chair, while Penny searched everywhere all the time calling her name. Koko then rushed out laughing. Koko imitated Penny on the telephone, screwing up her face and opening and closing her mouth. She kissed her dolls, tickled Penny's ears and cuddled her. When it was time for Penny to leave, the young gorilla clung on as if its very life depended on it. Going through Penny's mind was the thought that Koko's life just might depend on it. The separation, and return to the gorilla grotto, could kill her.

Penny fought for Koko and won. Stanford University agreed to house Koko and her trailer in its grounds and the zoo agreed to her being moved to a site closer to a potential mate called Kong at Marine World, Redwood City. On 19 September 1974, Koko moved house again.

The young gorilla arrived at the university before the trailer. She got increasingly agitated about being away for so long and continually signed 'go home'. As before, Koko had nightmares, and Penny had to stay with her every night for a month. But at least the experiment could continue.

Koko's mistakes continued to intrigue her teachers. She learned, for instance, the word 'straw' when shown a drinking straw, and then proceeded to use 'straw' for any straw-shaped object from plastic tubing to a car aerial. She also associated 'grass' with the colour green so a picture of a green pig became 'grass pig' and lettuce was signed as 'lettuce grass'. But she had also learned a good-sized vocabulary and was starting to beat Washoe. After three years of training Washoe had 85 signs, while Koko had acquired 127.

Kong proved to be an unsuitable mate and eventually the ownership of Koko came up for discussion. The new director at the zoo said that Penny could keep Koko if another female was found to replace her. Two young wild-born gorillas from Cameroon were for sale from an animal dealer in Vienna. Their parents had been killed and eaten by locals, was the official line, and they were orphans rescued from the rain-forest. Penny had qualms about purchasing the babies. There was no way to verify the story. It did, however, present her with the prospect of a young female for the zoo and a young male who could be companion to Koko.

Money was raised for the purchase with the help of the local and national press. On 9 September 1976 King Kong and BB (short for Brigitte Bardot) arrived, but the journey was too tough for the two baby gorillas. BB died. King Kong survived and was renamed Michael. The zoo, under pressure from the mayor of San Francisco, relented and did not insist on a replacement for Koko. In the summer of 1977 Koko was bought by the Gorilla Foundation – a non-profit organization set up by Penny Paterson, her companion Ron Cohn, and Barbara Hillier, who had been with Project Koko right from the start. It is entrusted with the welfare of Koko and Michael, and contributes both to research for people suffering communication abnormalities and to the study of gorillas in the wild.

The two gorillas lived separately but met every day. They would sign 'come' to each other through the fence and would ask if they could be let out to play together. In typical gorilla style they would have a rough and tumble in which Koko would sign 'chase' to Michael. Michael even learned a few signs that his teachers had not taught him. He had been watching Koko. Koko, for example, invented her own sign for tickle, in addition to the ASL one, which Michael could only have learned from her.

Their time at Stanford was a happy one. On mild Sunday mornings, Koko was taken for walks in the campus grounds. Students and Penny's weekend visitors stayed at a respectful distance, except on one occasion when one joined in a game of ball and hadn't realized that Koko always keeps possession of the ball. He ran off with it and got a bite on the behind for his trouble. Penny scolded Koko and later asked her why she bit the man. 'Him ball bad' signed Koko.

And while playing a game of chase with Eugene Linden (who co-authored *The Education of Koko* – a book about

Project Koko) Koko showed that she was capable of telling lies. During the romp, she bit Eugene on the hand, not severely but enough for Penny to demand what she had done.

'Not teeth,' replied Koko.

'Koko, you lied,' said Penny.

'Bad again Koko bad again,' signed the young gorilla, feeling suddenly very sorry for herself.

Another time Koko was breaking plastic spoons, much to her teacher's annoyance. She ignored any request to stop. Eventually, the teacher said 'Good, break them.' Koko stopped immediately and kissed them instead.

At first, about half of Koko's vocabulary had consisted of nouns, but by the age of seven nouns made up about two-thirds of the signs she had learned. This is similar to the vocabulary acquired by hearing children, deaf children and Washoe. The make-up of the rest of the vocabulary, though, was more like that of a deaf child. Definite and indefinite articles, the conjunction 'and', and auxiliary verbs like 'be', 'have' and 'could' were missing.

She had, however, a remarkable realization of the passage of time. Penny was once delayed in getting to the trailer to relieve the afternoon's teacher. She was 45 minutes late. Not long after the teacher's usual leaving time, Koko looked her in the eye and asked 'Time bye you'. The teacher signed 'What' and Koko again signed 'Time bye good bye'.

Koko's vocabulary differed in one major way from that of children, whether deaf or those with normal hearing, and that was in the use of 'who, what, where, when and why'. She answered questions that started with 'who, what and why', and on a couple of occasions used the signs for these words herself. One time, for example, Koko was asked to 'Open your mouth and close your eyes'. She was not unexpectedly suspicious and signed 'why'.

She differed from a deaf child in being able to hear human speech, and she demonstrated on occasions that she could also understand it. Koko had been exposed to spoken English before she was taught ASL. Her own name and the word 'no' would have been used on many occasions when she was in the home of the Children's Zoo director's family. Later in her education she surprised Penny, who was transcribing a sound tape, by acting out the words. The voice on the tape mentioned a broken spoon and Koko responded by breaking a spoon just after the words were heard. On another occasion a visitor had asked Penny what the sign for 'good' was. Before Penny could answer, Koko was producing the sign for the visitor. But, although she heard the words and made the appropriate actions, did she understand the words? Penny decided to find out.

The test chosen was the 'Assessment of Children's Language Comprehension' and Koko was tested under three different conditions – sign only and voice only, which were carried out blind by the teachers and recorded on videotape so that cueing was avoided, and sign-and-voice combined. Koko was asked to point at 40 different cards containing pictures of objects, such as a bird on a house, or depicting phrases, such as 'point to the broken boat on the table'. Koko did tolerably well – about as well as an educationally handicapped child. Significantly, she did as well with voice only tests as sign only tests, and did much better when the signs and voice were combined. Was her inadequacy due to some conceptual difficulty or was she just bored with the tests? The more difficult tests were carried out with the same degree of accuracy or inaccuracy as the simpler tests, indicating that comprehension and conception might not be the reason for the relatively poor result; poor, that is, compared to a normal child's.

In another study Penny analysed the answers Koko gave to various sets of questions when she was six years old. She was not looking so much for accuracy as for the appropriateness of the reply. An answer might be grammatically correct, for example, but totally inappropriate to the question. If Koko was asked to identify the colour of a blue towel and she replied 'red', the answer was incorrect and therefore inappropriate, even though grammatically in the right ball-park.

Koko was floored by 'what happened' or 'why' questions, but after the month-long study containing 427 questions was analysed, it was found that she had been grammatically correct about 83 per cent of the time. This six-year-old gorilla had performed better than two- to four-year-old children. Washoe had done nearly as well in the same kind of analysis.

After school one of Koko's favourite activites was a ride in the car. She always wanted to go to the same place – the soda machine. Koko knew the location of every machine on the campus and when asked where she would like to drive, would inevitably point to the nearest soda machine. At Ventura Hall, which was home to the Institute for Mathematical Study in Social Sciences, the computer programmers and statisticians were quite used to seeing a young gorilla walk to the soda machine in the back hallway, put in the money, take a can of lemonade, and walk nonchalantly back to a waiting Datsun car.

Some time later Koko, Michael, Penny and Ron moved to an old farm in the country at Woodside. Inside a special series of fences, barriers and doors Koko and Michael have a daily routine of lessons, exercise, play and mealtimes. The day starts at about 8am with breakfast and room-cleaning. Koko cleans her own room. She enjoys playing with Michael, which may end with a mutual grooming session, and then Michael is taken to his own quarters for ASL lessons with his own teacher, where he has been learning an ASL vocabulary to match Koko's.

Lunch is at 1 pm and dinner at 4.30. Bedtime is at 7.00. Koko has three rugs in a motorcycle tyre – she prefers to build a nest rather than sleep in a human-made bed. She has also made some interesting observations on the subject of death:

Teacher: Where do gorillas go when they die?
Koko: Comfortable hole, bye
Teacher: When do gorillas die?

Koko: Trouble old. And tigers
Teacher: Don't like tigers?
Koko: Frown bad red
Teacher: I like tigers
Koko: Tiger nail rough
Teacher: What animals do you like?
Koko: Gorilla love

But, next to gorillas such as Mike, and possibly 'honorary gorillas' – Penny, Ron, and Koko's oldest friend and teacher Barbara – Koko has a penchant for cats. Her favourite books were 'The Three Little Kittens' and 'Puss in Boots' and it came as no surprise, just before Koko's 13th birthday, that she told Penny she wanted a cat. Penny bought her a toy cat, but Koko was not at all happy to be fobbed off with that and sulked. She was now, after all, a pretty bright and eloquent gorilla with over 500 signs in regular use and a further 500 in reserve. Then a litter of three kittens was brought to the farm at Woodside. They had been abandoned at birth and had been wet-nursed by, of all things, a Cairn terrier.

Koko met the kittens, signed 'Love that', and picked out one in particular, one without a tail, which she named Little Ball. Koko was very gentle with Little Ball, despite being bitten and scratched. She treated the tiny kitten as a child would a pet, dressing it in nappies and hats, nuzzling it, and signing 'tickle'.

Sadly, Little Ball was killed, run over by a car. After Penny told Koko the sad news, Koko said nothing at first and then, after Penny had left, Koko cried. The story, though, has a happy ending. On hearing the news, a local cat breeder presented Koko with a brand new kitten, a red mackerel-tabby Manx. When it arrived at Woodside, Koko twirled with happiness.

Counters, Keyboards and Computers

Having eliminated the teaching of spoken English as an unsuitable method of talking to apes, and having established sign language as a means of exploring chimpanzee and gorilla minds, some researchers looked for other means by which people and apes could communicate – ways in which we could probe the intellect of non-human primates. Besides speech and sign languages, people use writing to express themselves and to pass on information and ideas. Clearly, it would be difficult to teach apes to write (although some have established themselves as 'accomplished' artists); but what if the 'words' were not written but represented in a way that an ape could manipulate? Then, a chimpanzee or gorilla could be taught to 'read and write'. And, by linking those representational characters to a simple keyboard, an ape could be taught to use a computer. Take that a stage further by adding a voice synthesizer, and the ape could, in theory, begin actually to converse in human vocal language. The prospects were exciting.

One of the earliest experiments was devised by David and Ann Premack at the University of Pennsylvania. The system used was not an electronic computer or a voice synthesizer but a mechanical interface using pieces of plastic. Each plastic token, with a different colour, shape or texture, represented a word (somewhat like characters represent concepts in Chinese). The pieces were backed with a magnetic strip, so they could be arranged in lines or 'sentences' on a metal board.

The pupil confronted with the task of making something of this experiment was Sarah, a wild-born chimpanzee who was about six years of age when she started with the project. A chimpanzee was chosen because the system was a visual one and chimpanzees have a visual capability similar to that of humans. In reality, Sarah was to be invited to communicate by a form of writing – not with a pen, but with representational pieces of plastic. Unlike her chimpanzee and gorilla cousins Washoe and Koko, who had to remember what they had learned, Sarah did not have to learn the 'words'; they were always in front of her. On the other hand, she was limited in what she was able to achieve. She was allowed to use only the tokens determined by the experimenters and was thus denied much of the interesting inventiveness and spontaneity shown by the others. Nevertheless, Sarah was given a vocabulary of about 130 'words' and was taught to 'read and write' instead of speaking. With this ability, it was thought she could explore the more logical properties of language and demonstrate the use of grammar rather than acquire a large vocabulary as have Washoe and Koko.

The training programme was much like that used to teach sharks, rats or pigeons to push pads, move bars or peck at keys. Sarah was introduced to an object and then by a series of simple steps began to associate it with a particular piece of coloured plastic. She was given an apple, for example, and first allowed to eat it. Then, she had to learn that in order to get another apple she must use the correct apple token and place it on the board. The apple token, like the other tokens, did not resemble the object it represented. The choice of colour or shape was quite arbitrary. Indeed, the apple symbol was a blue triangle.

The next stage was to introduce a new object and a new token, say a banana. If Sarah wanted a banana – represented by a pink square – rather than an apple, she had to place the banana token on the board. She was also introduced to the notion that her teachers had names and that each teacher was represented by his or her own particular token.

Having acquired a few 'nouns' or 'open' words, she was introduced to the use of 'verbs' or 'pivot' words. The word 'give', for example, was a useful action. In order to get the fruit reward, Sarah had to use 'give' with the teacher's name and thereby build up simple two word combinations. She also learned words like 'is', 'take', 'insert', and 'wash'.

Finally, she was presented with a token that represented 'Sarah' herself. At first she was required to put the token at the end of sequence. If she wanted the fruit, she had to use the sequence teacher-fruit-Sarah.

She learned quickly and went on to more complicated

'sentences'. With a bribe of chocolate she could be made to give up her fruit with a combination such as Sarah-give-fruit-teacher and even teacher-give-fruit-other teacher.

Amongst her 130 words, Sarah:
– learned how to 'write' with the negative adverb, thereby turning the meaning of sentences around;
– discovered the use of the questions 'who', 'what' and 'why'?
– acknowledged the concept of an object or a person, including herself, as having a name, which she recognized by the symbol 'is the name of';
– recognized size, shape and colour, first with words like 'red', 'yellow' and 'small' and 'large', and later suggesting more complicated relationships such as 'colour of' and 'size of';
– used prepositions and expressed conditions, such as 'if-then';
– constructed organized sentences.

In fact, her sentence construction was interesting. Sarah could understand the sentence Sarah-insert-banana-bucket-Sarah-insert-apple-dish and could equally well respond to or 'write' the shortened version Sarah-insert-banana-bucket-apple-dish; which meant that she was asked to put the bananas in the bucket and the apples in the dish.

In 1972, David and Ann Premack wrote in *Scientific American* that Sarah the chimpanzee had learned a simple language which included some of the characteristics of natural language. They had kept her learning programme as simple as possible, in a series of easy-to-learn steps. And having achieved what she did, they thought that Sarah's performance compared favourably with the language abilities of a two-year-old human child.

Sarah's experiment, like that of Washoe, was very time-consuming. Teachers had to be present all the time if any sense was to be made of Sarah's progress. The alternative was to create some semi-automatic system; in short, to replace human involvement partly with a computer, with the additional advantage that the experiment would reflect less of the researchers' personalities and be more objective.

A system was devised by Duane Rumbaugh and his colleagues at the Yerkes Regional Primate Center, Emory University, in Atlanta, Georgia. It consisted of a PDP-8 computer with two consoles, each containing 25 keys. On each key was a white geometric symbol, derived from nine basic shapes, used alone or in various combinations. Each symbol had a background, derived from three colours, and again placed singly or in different colour combinations.

If the object or action represented by the key was available for use, the symbol and its background was back-lit with a soft light. If the key was depressed, the light became brighter. Simultaneously, a picture of the symbol on the key appeared on one of several screens above the consoles. When a series of keys were pressed, the facsimiles appeared in a row in the order they were pressed, just like words in a sentence. The keys also had to be pressed in a particular sequence: the grammar. If the action or object was not available, no light was shown. One of the consoles was reserved for the use of teachers only and could be operated for teacher-pupil conversations.

The researchers gave the symbols or lexigrams and the order in which they could be used the name 'Yerkish'. They had devised their own special vocabulary with its associated grammatical rules. In effect, they had created an artificial language.

The pupil in this experiment was another chimpanzee, a two-and-a-half-year-old female by the name of Lana. In her first six months with the project she became familiar with and learned how to use the equipment. A variety of rewards were dispensed when she performed her tasks correctly. She could ask for food, toys, something to drink, music and even movies. She could request the window to be opened or a teacher to come in (when one was available) as long as the lexigram was softly back-lit.

Lana's first sessions were simple. Just as Sarah had come to associate one plastic token with an object, Lana was introduced to an available reward by pressing the appropriate key. Then, she was slowly introduced to the grammar of 'Yerkish'. Each request for a reward had to be prepositioned by pressing the key for 'please' and ended with the key for 'full stop'. When she pressed 'full stop', the computer analysed her performance and if she pressed the keys in the correct order, a tone sounded and she received her reward. If she made a mistake and the keys were pressed in the wrong order, the console and screens were cleared and Lana could start again.

Like Washoe, Koko and Sarah, Lana was fast to learn, and even discovered that, after making a mistake half way through a sequence, she could start again more quickly by pressing the 'full stop' key. She literally erased the sentence on which she was working before she had finished it.

The main interest, however, was the way in which Lana was able to read the word characters on the screen, recognize the beginning of a sentence, discriminate between valid and invalid sentence beginnings, and complete an incomplete sentence.

In one of the first tests, Lana was presented with a valid beginning to a sentence and six invalid ones. The valid beginning was 'please-machine-give'. She could then erase the beginnings or complete them using a selection of valid and invalid symbols. 'Juice' or 'banana' were valid words with which to complete a sentence because they could be used after the word 'give'. 'Music' and 'movie', on the other hand, were invalid words to use with 'give' because, according to the 'Yerkish' rules of grammar, they could only follow 'make'. Lana was 88 per cent correct in recognizing the invalid beginnings (such as please-object-give) and 100 per cent correct in spotting valid beginnings, (those that started please-machine-give).

In another test 'make' was substituted for 'give' and Lana was correct no less than 86 per cent of the time. And, in a third test, where only valid beginnings were presented but each contained various numbers of words, e.g. (1) please, (2) please-machine, (3) please-machine-give, (4) please-machine-give-piece, (5) please-machine-give-piece-of, Lana varied between 70 and 100 per cent success. The researchers concluded that Lana was able to recognize the principles of 'Yerkish' grammar and was able to read and perceive word order. She could tell the difference between valid and invalid sentence beginnings, even when the sentences were incomplete.

The work at Yerks has been taken further. Lana was joined by Sherman and Austin, and Duane Rumbaugh was joined by Sue Savage-Rumbaugh and their assistants S T Smith and J Lawson. They were out to prove that anthropoid apes can use the symbols they have learned as abstract representations. Their results were published in the learned journal *Science* in 1980.

The experiment was carried out in a series of stages and once again it started with a training period followed by a testing period. The first task was to sort out six items – an orange, a key, a piece of bread, a stick, some beancake, and several coins – into two bins. The food was to be placed in one bin, while the non-foods or 'tools' were to go in the other. Once they could sort the food from the non-food, the chimpanzees were asked to use the lexigrams for 'food' and 'tool' to label each item as it was presented by their teacher.

The three pupils were adept at sorting and using the lexigrams, but the researchers were faced with a question: Had the chimpanzees learned simply to associate an item with a bin or lexigram, or had they realized something more general – that one bin was for things that could be eaten by chimpanzees and the other was for non-foods?

To answer the question the researchers conducted a blind trial. The chimpanzees performed in their room with the computer and the teachers remained outside with the object with which they were to be tested, in this case five 'tools' and five foods. Each chimpanzee went in to the teacher and was shown the test object. The chimpanzee then returned to the computer and was expected to push one of two lexigram keys, either 'tool' or 'food'. At that moment, the chimpanzee could not see the object and the experimenter could not see which key the chimpanzee had hit.

The results were interesting. Austin went to the top of the class by correctly categorizing all the objects. Sherman got one wrong, but was excused on the grounds that the object was a bath sponge and he was notorious for eating sponges. Poor Lana, who had shown such promise in the earlier experiments, flunked completely. In the first test she only got three right and in a later test only one.

The test showed, the researchers thought, that Austin and Sherman had appreciated the concepts of food items and non-food or 'tool' items. Lana had learned only the specific sorting responses. In more experiments, however, even Lana got the tests right. It was thought that she had a problem with the concepts of 'food' and 'tool' at first because she could not represent her concepts as symbols.

Austin and Sherman were taken a stage further. Instead of seeing the objects themselves, they were presented with photographs of the objects. Again, the two star pupils passed with flying colours: they correctly categorized every item.

By this stage the two chimpanzees were correctly categorizing objects or photographs of objects. Next they were to be asked to sort out lexigrams that represent objects – not the objects presented in the previous tests but objects in their normal lexigram vocabulary. This was an important test. To succeed, the chimpanzee had to recognize which item was represented by the lexigram, decide whether it was 'food' or 'tool', remember which was the 'tool' lexigram and which was the one for 'food', find the correct lexigram, and finally press the key.

First, they were trained with six of the original test objects to give them the idea, and then they were asked to sort out 17 lexigrams from their lexigram vocabulary. Again, Austin was 100 per cent correct and Sherman made a single mistake.

The chimpanzees had been successful in passing the tests. The researchers felt it was safe to say that they were able to categorize objects according to their function. Lana's earlier failure to do so but later success suggested that chimpanzees can categorize objects without the need to organize that knowledge into symbols. The difference between Lana and the other chimpanzees was that they were trained to talk to each other and had been trained to use tools to obtain food. It was this training, which Lana lacked, that helped Austin and Sherman to consider 'food' and 'tool' as general, representational categories. This also gave them the edge when presented with novel foods and tools. They were able to categorize them with considerable ease. The amazing conclusion arising from these experiments is that a chimpanzee – a non-human primate – can understand that abstract symbols represent not only particular objects but also the intrinsic properties of those objects.

Next, Austin and Sherman were introduced to the concept of sharing. To do so, they had to talk to each other via the computer. Sherman had to press the correct lexigram to tell Austin where the food was hidden. After months of training, they succeeded.

And if all that should not encourage a reassessment of the uniqueness of man's place in nature, there was more to come. The idea of using computers in ape language studies has been taken one stage further in California. The work is almost a return to the original plan to teach an animal actually to speak English. And centre-stage once again is Koko the gorilla, with Penny Patterson.

Koko's excursions to the soda machines of Stanford's Institute for Mathematical Studies in Social Sciences were not entirely without scientific interest. Professor Stuppes and his colleagues at the Institute took an interest in Penny Patterson's work and designed a keyboard-computer system that enables Koko to use vocal words. By pressing the appropriate button, the gorilla can activate a synthesized voice, while the computer itself records the activity. Penny

Patterson had noticed that Koko responded to vocal words as well as sign language and thought that it would be an interesting experiment to find out what Koko might 'speak' back, if she were able to. The keyboard has the usual 46 keys with letters of the alphabet and numbers. Koko's keyboard differs in that each key is painted, in one of ten different colours, with a geometric sign. Each key represents an object, an action or a feeling.

Koko is not a Pitman's speed-champion but a plodding one finger 'jabber'. She always uses the index finger of her right hand for the keyboard, leaving the left hand free for signing. As the words are duplicates of the ones for which she knows the ASL signs, she can speak with the speech synthesizer and make the sign gesture at the same time. An ambidextrous and bilingual gorilla certainly makes a mockery of assertions made in the late 1950s that gorillas were intellectually inferior to chimpanzees.

Objects are placed in front of Koko and she is asked to describe them or comment on them. An apple, for example, may prompt Koko to push the keys and make the signs for 'want', 'apple', 'eat'. The synthesized female voice will follow the action.

The idea behind the experiment is to discover a gorilla's sense of spoken-word order and to see if Koko can construct syntactically significant sentences. The work goes on.

Nim the Terrible

Nim Chimpsky was his name, and after four years of training with more than 60 volunteer teachers, at an estimated expenditure of $250,000, he was to throw cold water over the ape-language experiments and support his namesake Naom Chomsky's notion that language is innate (we have a 'language organ' in our brain) and is unique to man.

Nim's project started in December 1973 at Columbia University in New York, although his life started at the Oklahoma Institute for Primate Studies where he was born.

At just one week old, the baby chimpanzee was flown to a Manhattan home where he was to live the first few weeks of his life. His mentor was Herb Terrace of the department of psychology (the same Herb Terrace who had suggested the role of animal interpreter for a singing ape), and it was Professor Terrace's intention to 'collect and analyse a large corpus of a chimpanzee's sign combinations'.

Nim eventually went to live in a Riverdale mansion in the Bronx with three of his teachers, and went to school in a specially designed classroom at the university. Nim was the only pupil, and he was to learn and use ASL.

At Riverdale Nim had his own bedroom and was pot-trained by the age of two and a half. He was treated in much the same way as a child might be, joining in the day-to-day activities, such as dressing, helping with the cooking and tidying the house. It was not a desire to turn Nim into 'a middle-class chimpanzee', as Herb Terrace put it, but to provide him with a stimulating environment in which he had every opportunity to learn new signs associated with everyday life.

At school Nim had six teachers at a time. He would wear them out after five or six hours of intensive work. And for every hour of teaching and testing, there was at least another hour of preparation time. Project Nim was very demanding. The experiment was also demanding on the young chimpanzee. With the constantly changing teaching staff, it became increasingly difficult as the experiment progressed for the teachers to keep Nim's attention. Nevertheless, during the 44 months the project ran, Nim learned 125 ASL signs and he could combine them in various ways, just as Washoe had done.

Analyzing the combinations made during two years, the researchers recognized that Nim had made 1378 novel sign combinations of two or more signs, and had used them about 20,000 times. But were they primitive sentences? At first, Herb Terrace thought they were.

Certain of Nim's signs were always used in the same

position in the two-word 'sentence'. In about 85 per cent of cases, the sign for 'more' preceded the sign for whatever Nim wanted more of, as in: 'more banana' or 'more tickle' (chimpanzees *do* like to be tickled). Likewise, 'give' was signed first in 78 per cent of combinations, such as 'give banana' or 'give apple' (they also like their food). He also used 'pivot' words or verbs, such as 'hug' or 'tickle', with the signs for himself, such as 'me' and 'Nim'. The verb came first in 93 per cent of combinations.

This regularity of word order was interesting, thought Terrace, but it did not demonstrate that the chimpanzee was learning or using any grammatical rules. It was not random, however. There was certainly something in it. Nim could have been imitating his teachers. On the other side of the coin, it was unlikely that the combination 'tickle me' would be first signed by a teacher. There was a far greater motivation for Nim to instigate *that* combination. And 'give' would have been used with the sign for whatever object Nim wanted long before there was any desire to pass the object the other way, i.e. *to* the teacher. Herb Terrace gradually persuaded himself that the two-word and more-than-two-word combinations were grammatically significant and not unlike the first simple sentences learned by a young human child.

Like the other signing apes, Nim had his idiosyncrasies. If annoyed by somebody to the point where he wanted to bite them, he pulled back his teeth, put his hands above his head, and his hair stood on end. But, before he actually attacked, he made the signs for 'angry' and 'bite' and the rage gradually calmed down. In effect, by using the ASL signs, Nim had expressed the emotion of anger through symbols, another human-like attribute.

He also learned how to cheat his teachers. The sign for 'dirty' was used when Nim wanted to go to the toilet, and he quickly realized that when he gave the sign all the teaching stopped and he was hauled off rapidly to the toilet. Many times he signed 'dirty' but really didn't want to go. Professor

Terrace noticed he made the sign with a grin on his face, and the wily chimp avoided eye-contact. Nim had been winding his teachers up; he had manipulated their behaviour to his own advantage by using sign language.

Having reached this stage in the experiment, it had to be terminated. The project simply ran out of funds and Nim was returned to the Oklahoma Institute for Primate Studies. Herb Terrace and his research team suddenly had a lot of time on their hands; they started to look more closely at their results, and their interpretation of the data began to change.

Professor Terrace was concerned that, although the size of Nim's vocabulary increased, the number of words in his word combinations did not increase significantly during the course of the experiment. The average was between 1.1 and 1.6 words, a score more likely to be demonstrated by a human child who had just started to learn how to combine words, rather than by one who had been learning for almost four years. In addition, Nim's long utterances – such as his record-breaking 16-sign give-orange-me-give-eat-orange-me-eat-orange-give-me-eat-orange-give-me-you – did not build any extra meaning into the sentence as a child would do.

There was more to come. The researchers looked at the video tapes they had made during the teaching and testing sessions. In about 88 per cent of the time conversations were initiated by the teacher, not Nim. A child would have started up a far higher proportion of conversations. Children are more spontaneous.

Also a child imitates its parents less and less as it gets older. At 21 months, a child copies about 20 per cent of the time. At three years, it has stopped direct copying altogether. Nim, on the other hand, imitated his teachers in 38 per cent of utterances when he was 26 months old, and curiously increased the amount of copying to 54 per cent when 44 months. As children stop copying their parents, they tend to expand on what their parents have said. Beyond a certain age a child's sentences become longer and more structured. Nim did not do that.

The chimpanzee also interrupted his teachers often. As they get older, children do not do that either. They gradually learn that conversations are built up by one speaker adding information or a comment to the previous one, and, by observing hidden word cues, inflections in the speech, or some element of body language, they take turns in speaking. Nim, according to Professor Terrace, very rarely added to a conversation and certainly did not take his turn. The researchers were beginning to find significant differences between the way Nim and a human child used language.

Terrace felt that Nim's signing appeared 'to have the sole function of requesting various rewards that can be obtained only by signing'. All the signing activity is simply an irritating and meaningless way to get to the reward at the end of the session. The teacher will try to engage the chimpanzee in a conversation about some activity, like discussing the colour of an orange, and all the chimp wants to do is to get through the process as quickly as it can to eat the orange. No wonder, thought Terrace, the chimp interrupts so often.

This is further illustrated by the signing of 'cat'. Nim learned how to sign 'cat' and also learned to use it in order to play with his pet cat. But was this just a trick that Nim used to get to the cat? Did the sign have any other functions? Terrace felt it did not. Nim did not ask 'where is the cat?' or say 'there is a white cat over there'. In other words he did not exchange information. During the supposed conversation, all the chimp was doing was to imitate his teacher and throw in a few extra words that to the human viewer looked like grammatically interesting utterances. In effect, Terrace concluded, Nim was simply making a demand, not holding a conversation, and he wanted his demand realized as fast as he could possibly make it. Terrace was doubtful if Nim had demonstrated any signs of grammatical competence. Shock waves reverberated through the ape-language community.

Washoe's achievements now came under close scrutiny,

and the mud-slinging began. One major criticism was that scientists engaged in sign-language experiments tend to over-interpret the sometimes ill-defined signs the animals are making. Like doting parents, they see complexity in the simplest events. Re-examination of the videotapes taken at signing sessions, the critics say, sometimes shows what appear to be long and complicated conversations but which are, in reality, a confused series of gestures that the human teacher sorts into neat sign combinations or 'sentences'. These criticisms are understandably vague, because published conversations are often edited, with the added complication of observer bias.

Fundamental to the argument was whether the apes had even been taught ASL. Two of Herb Terrace's students, Mark Seidenberg and Laura Pettito, pointed out that fluent users of ASL arrange their signs in an order that in no way resembles the grammar of spoken English. ASL has a structure and rules of its own and is a language in its own right. The ape researchers, they thought, take their vocabulary from ASL but loosely impose the grammatical rules of English. The 'language' being taught to the apes is a kind of pidgin-sign-English. It is starved of the grammatical features both of spoken [illegible], such as differences in sounds, the use of suffixes, and [illegible] order, and of ASL, which depends on the way signs are performed. Seidenberg and Pettito thought that the form being taught to apes was not a real language, but a manual code for English.

Another feature of ASL is that, as users become more proficient, they move away from the rules of English grammar and gradually move towards the rules that govern ASL usage. Unfortunately, linguists have not yet set down all the rules of syntax and semantics associated with ASL, and it is not taught formally. Inevitably, the apes, like their deaf human counterparts, begin to show ASL-type signing, and English word order may not be appropriate. In this situation it would be difficult for a non-fluent ASL signer to evaluate the sign sequences being used by the apes.

The significance of all this, of course, is governed by what is considered to be language. Is early child speech, with which ape sign-language achievements have often been compared, included as language, or must we wait for adult usage before the term applies? Can we assume that young children will eventually speak the language of their parents and that their early utterances are simply an interim phase? With apes, that assumption cannot be made. We can evaluate only what we see. Furthermore, how do we interpret what we see?

There is a good human analogy which illustrates the problem. A young English schoolboy is on holiday in France but has not learned the language. Children (as I have discovered with my own) do not need language to join in games, and so it is no surprise when his father finds him playing football with the local boys. But he is impressed when the boy, spotting a gap in the opposing team's defence, shouts 'Ici, ici', and the ball is passed to him and he scores a goal.

After the game has finished the father, – also non-French speaking – asks his son what 'ici' means.

'I really don't know,' the boy replies, 'but every time I shout it, I get the ball.'

The boy had mimicked the other boys, used the right word, and it had achieved what he wanted, but he had no idea what it meant. It is the difference between 'knowing how', as G. Ryle put it in *The Concept of Mind*, and 'knowing that'. Chimpanzees clearly 'know how', but how do *we* know whether a chimpanzee is 'knowing that'? We can simply ask the boy about the event, but we have not yet devised a way to ask a chimp. In short, do the gorillas and chimpanzees comprehend what they are saying and doing or are they simply conditioned to say it and do it? And how do we find out?

This problem of interrogation is illustrated by another example. How do we know, for instance, when invention is true invention? Washoe once saw a swan for the first time and signed 'water-bird'. Was she giving the bird a name which she had invented herself or was she simply acknowledging that she could see a bird and that she could also see water? Was this apparently novel word just an accident? Similarly, Nim combined the sign for 'banana' with a host of unexpected words, such as 'banana toothbrush'. Was the toothbrush shaped like a banana, the colour of a banana, or, as Adrian Desmond suggests in his book *The Ape's Reflexion*, did the chimpanzee want simply to eat a banana and clean his teeth with a toothbrush? Herb Terrace thinks the latter explanation unlikely because both objects were not in view at the same time, and Nim, unlike Ally (see below), rarely signed for objects he could not see.

Koko the gorilla invented some intriguing new words. She called a zebra 'white tiger', a face mask 'eye hat', a watermelon 'drink fruit', and Brazil nuts were labelled with the combination 'rock berry'. Had she rightly observed that watermelon was a fruit that was full of water, like a drink; that a Brazil nut was a small fruit (which she called berry) which is hard as a rock; that the mask was placed on the eyes like a hat on the head; and that zebras have stripes like tigers? Or was there some other stimulus for the combination – perhaps, word-play? And, again, how do we find out? One of the criticisms of the ape-language work is that these novel combinations have not been well documented and are therefore difficult to analyse. However, C A Ristau and D Dobbins, in a paper presented at the 1981 Darlem Conference, *Animal Mind – Human Mind*, suggested that researchers might be able to distinguish between the novel use of words and word-play by watching for facial expressions and other aspects of behaviour during the signing session.

While the psychologists and philosophers work all this out, the debate goes on. Those on the defence, who are out to promote ape-language studies, have been fighting back.

The Gardners, who worked initially with Washoe, suggest that it is invalid to compare the way in which young chimpanzees use sign language with fluent human adults. They also question the relevance of frame-by-frame analysis of the video tapes taken at signing sessions, from which critics have suggested cueing is taking place. Words in ASL, which is a visual language, can be held for longer than spoken words. Overlapping spoken words are difficult to hear and so we have developed a rule that speakers tend to follow one another in a conversation. It is interesting, though, that overlap occurs so often in conversations on the telephone, when visual follow-on cues are missing. For the deaf, overlap is not important. Two deaf people, having a conversation using ASL, often have their signs overlapping. The same is true of the chimpanzee and gorilla conversations. The Gardners therefore challenge the significance of overlap and deny that they are victims, albeit unconsciously, of the Clever Hans syndrome.

Roger Fouts draws attention to the fact that his chimpanzees can refer to objects that are not actually there. In one experiment, Nim's brother Ally, at the Oklahoma Primate Center, watched as several objects were hidden in a screened-off part of the chimpanzee's test room. He was then taken to another part of the room and interrogated by another teacher who had not seen where the objects were hidden. The teacher asked Ally about the locations of the objects with questions such as 'where-ball?' Ally answered by signing 'ball-under-box' and so on. Out of 240 tests, he signed preposition-location relationships correctly 161 times. Ally had demonstrated a form of symbolism, in that the signs he used represented things that were not actually present. This is an accepted criterion of language.

David Premack, writing in his book *Gavagai! or the Future History of the Animal Language Controversy*, feels that the sign-language performances they have been witnessing are the products of learning, and give us some idea about the learning capacity of chimpanzees and gorillas. He acknow-ledges, however, that in human-language acquisition there is more going on than just learning. At first, children learn words and their meaning in a rather haphazard fashion. Chimpanzees do the same. Children then do something different, it is thought. They rearrange the various elements they have acquired in order to introduce some regularity. This shows itself as rules of syntax, and often there are understandable errors. A child might first use 'foots' instead of the correct plural 'feet', or 'begunned' instead of 'began'. This stage of language acquisition is thought to be the phase peculiar to the human brain. In computer terms, it is the 'hard-wired' process of rules selection that Naom Chomsky believes is denied to other animals: which brings us to thinking.

If an animal does not have language, as we understand language, can it be said to be thinking? As there is no evidence for natural rules selection in the sign-language experiments, the researchers must, as Premack puts it, 'simulate by a process of learning that which we ourselves cannot carry out by learning'. There is 'no recipe for language'. So, without language, no thinking.

Language, as D Dorner of Banberg University in West Germany points out, is a control system that regulates what would otherwise be disorderly thoughts. He suggested at the 1981 Darlem Conference:

> A sentence, as a sequence of words bound together by grammatical rules, is, technically speaking, a device for organizing these disorderly thoughts. The rules of language aid in this organization.

In this way, Dorner suggested, language allows an ordered flow of thinking which replaces the disorderly sequence of association, such as those one might experience in a dream. Language allows people to find out things for themselves by using elements of their memory independent of events taking place in the outside world. If apes are not able to combine words in meaningful ways, it is suggested that this severely restricts their ability to think.

What, then, if somebody proves that animals do think?

Guy Woodruff, at Temple University in Philadelphia, has a chimpanzee called Sarah who shows basic mathematical reasoning. She can judge whether one pair of objects is related in the same way as another pair. And researchers working with rats in a specially designed maze at Columbia University found that the animals could remember and think.

A rat was placed at the centre of a maze with eight identical arms. At the end of each arm was placed a food pellet. The rat had to collect all the food pellets and bring them to the centre. After a few runs, the rat soon learned to empty the arms without visiting the same arm twice. In successive runs it did not enter the arms in the same order, sometimes going for 5,6,8,7,1,3,2,4, while on another occasion visiting in the order 6,1,8,3,2,5,4,7 and so on. The animal, it was confirmed, was not following a scent trail or relying on any other external cues. It was remembering which arms it had already visited and thinking about which arms it had left to visit.

In another experiment, Herb Terrace at Columbia chose another creature to which we do not attribute a great intellect, namely, a pigeon. He put the pigeon in an experimental chamber with four response discs, the colour of which could be changed at any time. In order for the pigeon to obtain food, it had to respond and peck at the discs in a particular order, say,a red, green, blue and yellow. But, although the sequence order remained the same throughout, the positions of the colours were changed.

Herb Terrace thought this was a complicated thing for a pigeon to do. He likened it to a visit to the automatic cash dispenser at the bank, where the numbers were colour coded and the bank manager changed the position of the colours. Everytime you went to punch in your ID code, you would find the numbers and colours in different places on each visit. The pigeon is faced with the same problem. Red might have been in the upper left-hand corner yesterday, but it is no longer there today. The pigeon had to seek out the colour, and then each successive colour in the sequence. Put simply, it had to think.

He also discovered something else. Professor Terrace wanted to speed up the process of learning and found that if he introduced shapes such as triangles and squares, and mixed them up with the colours, the pigeons could learn faster. But this worked only in certain combinations. If the pigeons were presented with colour-shape-colour-shape-colour they did no better than previously, but given colour-colour-colour-shape-shape the birds learned a five-element sequence in a fifth of the time. What the birds were doing was a process known as 'chunking'.

'Chunking' is used by us as a memory aid. It is a way of reorganizing superficially unrelated items into higher-order units. Take for instance, the telephone number 747 1066. You can look it up in your address book or the telephone directory and say it over and over again until you have dialled it. A few moments later, you will have forgotten it. That is short-term memory. If, however, you wish to remember the number permanently, you might break it down into 'chunks' and maybe use the mnemonic approach – jumbo jets remind you of 747 and the date of the Battle of Hastings reminds you of 1066.

'Chunking' in people presupposes a large storehouse of knowledge that can be searched for suitable mnemonics. If a person wants to reorganize the new information in this way, then 'meaning', which has been stored away in auto-memory, is enlisted to help process the information. In human terms this implies the use of language, because meaning for people is embedded in symbols called words.

This, suggests Herb Terrace, throws up a new problem. If (1) these experiments are proof that animals *do* think, (2) it is accepted that thinking and language go together, and (3) the sign language experiments indicate that animals do not acquire language, then how do animals think without language?

The Gardners, Roger Fouts and Francine Patterson,

however, are convinced that language or something akin to language *is* involved, and they are concerned that the debate over what to call it is detracting from the phenomenon itself. David Premack suggests that the sign-language apes are 'more than mindless, yet less than linguistic'; but this preoccupation with the nature of language means that researchers are ignoring more interesting comparisons of cognition, the faculty of knowing and perceiving.

Indeed, what can be made of the occasions when chimpanzees sign to themselves without teachers present? Another trained chimpanzee, Lucy – the charge of Maurice Temerlin at the Institute for Primate Studies in Oklahoma – would sit and read a magazine and sign to herself some of the objects she saw in the pictures. She was not attempting to please anybody, not trying it on simply to get a reward. Was this simply a spill-over from the training sessions, a conditioned response which said 'there's a banana, I'll sign because I usually get a reward', or was the young chimp recognizing an interesting banana?

At a Scientific Meeting of the Zoological Society of London in 1980, Mary Midgeley, of Newcastle University, drew attention first to Temerlin's work and then to the fact that there seemed to be only two sides in the sign-language dispute: those who believe that apes 'do it just like children' and the others who say 'it's pure chance'.

'This simple antithesis,' she argued, 'is surely wrong. Apes obviously don't think like people. (If a child had been brought up by orang-utans would it climb trees like an orang-utan? No, but it would still climb trees a lot better than we do.) What interests me is to know what they actually *are* doing.'

In response to Mary Midgeley's remarks, Herb Terrace acknowledged that even Nim had demonstrated some behaviour for which he had no explanation and which was certainly worthy of further research. He recalled occasions when Nim signed spontaneously.

While driving home one day with Nim and Professor

Terrace, the bus driver stopped at traffic lights and took the opportunity to pour himself some coffee. Nim pointed to the driver and signed 'drink'. Nim had just had his fill of orangeade before leaving the classroom so he certainly was not thirsty. And, when they got home, Nim was offered more drinks but really did not want any more. So, why did he sign 'drink' in the bus? Was Nim signing in an informative way and not simply trying to influence his teachers to give him something he wanted?

Even if we accept the bottom line – that chimpanzees and gorillas using ASL, plastic tokens, or 'Yerkish' computers are manipulating their masters to give them just what they ask for – it offers intriguing insights into the minds of our close evolutionary cousins, the anthropoid apes.

It reminds me of a story told by Martin Orne at the Clever Hans conference in New York in 1980. He recalled a student prank. Lecturers, he reminded the audience, are flattered by students taking copious notes, which leaves them open to 'positive reinforcement'. The students manipulate the lecturer simply by taking notes only when he reaches a certain spot on the dais. A well co-ordinated class can soon have its lecturer teetering on the edge and about to fall off. The lecturer believes he is teaching the students, but, in reality, it is he who is being taught.

3

Conversations in the Wild

What Wild Animals Say

The 'sympathetic' approach to finding a window through which one can see what makes an animal tick is to study the animal in its natural habitat and attempt to find out what exactly it is saying and doing. In other words, instead of teaching animals one of our contrived 'languages' such as ASL, we try to understand theirs. One creature has even exploited our curiosity.

In Africa a small bird known as the honey guide has turned the tables on man and uses its calls to lure people to help *it* get a meal. Normally, the honey guide uses a special call to lead honey badgers (ratels) to the nests of bees. The badger breaks open the nest and eats the honey, while the honey guide is satisfied with the wax that makes up the comb. Local people realized that they could obtain honey just by following the honey guide and ratel. The birds eventually 'discovered' that people were just as good at finding bees' nests, and began to call to them for help. More recently, however they have given up calling for people: the local human population no longer relies on this source of honey – it's easier to get it from the supermarket.

The symbiotic (that is, mutually beneficial) relationship of the honey guide and the ratel is often quoted as an example of interspecies communication and cooperation. It works not because the ratel understands the actual meaning of the honey guide's call but because it has come to associate the sound with the reward of food.

Animal behaviour researchers, unlike the honey guide, are not concerned with food but with what other animals are saying. Their work is not easy, for animals communicate in all sorts of ways using many different channels. And any attempt we might want to make to strike up a conversation with those speaking at the extremes of human hearing or vision are going to be fraught with difficulty. But instead of attempting a conversation, most scientists just listen, trying to understand what is said by recording the sounds or watching the gestures and associating them with particular patterns of behaviour. Slowly, they are beginning to unravel the nature and meaning of the messages passing between individuals of other species.

One method of achieving this is to mimic an animal's calls, or play them recordings from a tape recorder, and note how they react. This technique is called 'playback' and it is likely that man has been using it since he started to hunt.

Playback for the Pot

Early man, foraging, scavenging and hunting on the African savannah (and also in the Middle East and southern Asia, if current research is anything to go by), probably began to notice the calls and songs of other creatures and learned that by imitating them he could lure the animals to come closer, perhaps within range of a primitive weapon, or at least to reveal their position. In a paper published in *Recorded Sound*, Jeffery Boswall and Robert Barton review the 'Human Imitation of Bird Sound' and reveal that this form of man-animal communication is common throughout the world.

Paul Henley, for instance, has described the way the Panare tribe in Venezuela entice birds within blowpipe range by mimicking their calls. In the Upper Negro region of north-eastern Brazil, Howard Reid found that the Maku Indians copy the call of the marbled wood-quail and several other forest birds. As the birds approach the calling hunters, they are despatched by bow and arrow.

Prehistoric man did not rely only on his own vocal abilities: he invented devices to help with the imitation. In New Mexico, excavations at archaeological sites have revealed bone flutes that almost certainly were used to call to wild turkeys. As recently as the beginning of the twentieth century, Pueblo Indians had been seen to use similar flutes to lure the birds.

Some of the devices were complicated inventions. In *The Natives of Sarawak and British North Borneo*, H Ling Roth describes a bird-call instrument which consists of a narrow tube into the top of which is inserted a long thin piece of bamboo. It was used by the Kadyans to attract emerald doves towards a hunter inside a hide made of forest leaves.

Hunters in hides were in evidence in Tanzania in the 1930s. In their book *Empire of the Snakes*, F G Carnochan and H C Adamson describe a particularly ingenious method of capturing starlings for the pot. Young boys learned to imitate the mobbing calls of the local starling population:

> The boys would build huts that looked like inverted ice-cream cones and big enough to hold several boys. The walls of these structures were made of heavy branches, the leaves of which hid the boys inside. The twigs of these branches would be heavily smeared with a bird-lime that was made from the milky syrup of the fig tree that stood on the village square. When several of these huts had been built where the edge of the bush framed the cultivated corn or yam fields, the boys entered them singly or in pairs, and began to imitate the peculiar strident sound of a starling that has been captured by a hawk. Eager to assist their captured comrade and hack the hawk to pieces in a concerted attack, flocks of starlings would fly toward the sound and swoop around the huts looking for the hawk. Every once in a while a

> starling would light on a hut and get tangled in the bird-lime on the twigs. Then a small black hand would shoot out through some opening and catch it by the legs, tail, or any seizable part, and drag it inside.

What these boys of the Wanyamwesi tribe, like the Maku, the Panare, and the Kadyan, had achieved was a simple but effective form of communication with another species. Communication is said to have taken place when transmitted information from a sender has influenced a listener's behaviour. The fact that the transmissions have been one-way, and wholly to the benefit of the sender, in no way invalidates them as communication. On the contrary, they fit with Professor Peter Slater's (University of St Andrews) view about the relative benefits to senders and receivers during communication, and that evolution favours the sender.

In most cases the hunters simply exploited normal territorial or courtship behaviour shown by the birds; they had realized that the calls or songs that they imitated were sufficient to stimulate the bird's curiosity or activate some innate behaviour, such as mobbing.

One researcher, Edward Armstrong, has even gone as far as to suggest that the prehistoric imitation of animal sounds as a hunting aid may have been a precursor of true language. Early man's utterances would have been limited, and the development of imitations would have enhanced the range of sounds that could be made. Also, the imitation of the sound of the prey would help to focus the attention and co-ordinate the actions of all those taking part in the hunt, thereby making it more likely to succeed.

Today hunters and shooters still imitate animals, particularly birds, to bring them in range of their guns;. sounds can vary from simple vocalizations, such as whistles, to electronic gadgets that make the sounds for you. In North America 'spishing' and 'squeaking' are two methods employed.

Spishing is made by forcing air through closed teeth. Woodpeckers, nuthatches, warblers and sparrows are apparently attracted to the noise; swallows are indifferent, however, and blackbirds fly away. Why there should be such diverse reactions to the same sounds is debatable. Squeaking is a high-pitched sound made by forcing air through pursed lips or by placing the lips on the back of the hand. It is thought to resemble the alarm call of a young bird and incites birds to mob a non-existent intruder.

Then there are the artificial whistles and calls, such as curlew, cuckoo or duck calls. French shooters attract larks and finches with a flat, circular device that you suck rather than blow. And pipes attached to hand- or foot-operated bellows, such as the French quail pipe or the Scotch duck call, are used to imitate birds with rapid songs or calls. 'Crake bones', which consist of two pieces of horse ribs that are rubbed together mechanically, are attractive to corncrakes. And resonating boxes, in which the cedar-wood lid, hinged at one end, is rasped against a piece of slate, is fatally attractive to snow geese. (It can also be fatal for the caller: on at least one occasion, the artificial call has been so convincing that the operator has been shot by a fellow hunter!)

At the top end of the scale are the electronic devices, the main one being the tape-recorder/player. Although not often used by hunters, the tape recorder came into its own when the pioneering wildlife film-makers wanted to entice animals in front of their lenses. The technique is still used. It has not only brought the birds of the world into your lounge via the TV set, but has also been an important scientific tool in the study of what animals are saying.

Playback for a Lost Whale

Humphrey took a wrong turn and found himself up a Californian creek, well and truly trapped. Humphrey was (and hopefully still is) a 12-metre-long, 45-tonne humpback whale, and on 10 October 1985 he was on his annual trek from the feeding grounds off Alaska to the breeding grounds in Hawaii when he took a left instead of a right and ended up in the Sacramento River, about 100km from the Pacific Ocean. Instead of turning about and rectifying his mistake, Humphrey decided to stay. People were understandably concerned for his safety: fresh water is less buoyant than sea water, making it more difficult for Humphrey to swim properly, and it was feared he might become exhausted. Fresh water might also have damaged his skin, causing it to peel off. If that had happened, he would have not survived for long.

So an operation was launched to return Humphrey to the sea. First, killer whale sounds were played to frighten him downstream (killer whales sometimes chase and kill humpbacks). Humphrey ignored them. Next they tried fire-crackers, but probably succeeded only in deafening him. Someone suggested an ancient Japanese method for driving away dolphins from fishing nets, and so people advanced towards Humphrey in a line of small boats, banging on gongs and pipes. They succeeded in moving him downstream past two river obstructions, including a drawbridge that had to be raised for two hours, causing traffic tail-backs for miles around. But it was all in vain: during the night, Humphrey sneaked back up the river.

Marine zoologists from California, Florida and Hawaii went into a telephonic huddle and, after a four-hour-long conference call, a strategy was devised. It was based on work by Joe Mobley, Adam Frankel and Louis Herman, who were then all at the Marine Mammal Research Laboratory in Hawaii, and had been studying the vocalizations made by humpback whales at their feeding grounds in Alaskan waters.

Humpbacks mainly feed alone, but occasionally groups of whales feed cooperatively. They appear to surround and swim under a school of tiny fish, and at the command of a single 'co-ordinating' whale, which makes a loud 'trumpeting' sound, they all rush to the surface with mouths agape and jostle at the surface as they scoop up the fish.

As part of some experiments on the whales' breeding grounds off Hawaii, the Hawaii research team decided to compare the reactions of whales exposed to playbacks of songs made on the breeding grounds with those presented with the 'feeding' sound. The results were dramatic. The whales generally ignored the winter breeding songs, but as soon as the summer 'feeding call' was played, they turned in their tracks and charged towards the sound source. Here, thought the researchers, was a means of enticing Humphrey back to sea.

Tapes of the 'feeding call' were sent from Hawaii to San Francisco and a ship named *Bootlegger* was made ready for sea, equipped with underwater playback facilities. A woman scientist went out in a rowing boat with a tape-recorder and hydrophone (underwater microphone) in order to monitor what went on.

She signalled that she was ready, and the call was played. Humphrey's response was immediate and took the researchers by surprise. He swung around and charged the *Bootlegger*. The woman rowed rapidly back to the mothership, scrambled on board, and the *Bootlegger* set off downstream, every now and then playing the 'feeding call'. Humphrey followed faithfully for 85km, and after seven hours later he was back to the Golden Gate at the entrance to the almost land-locked San Francisco Bay. And there, fortunately, he stayed for the night. Next day, the same procedure was followed and Humphrey was led into the Pacific. He has been seen every year since, alive and well and swimming off the Californian coast about 40km to the west of San Francisco.

Humphrey's story shows that humpback whales, like many animals, have their own communication systems, maybe even languages. If we could understand them, we would gain tremendous insights into the way they live and survive.

Playback for Science

Playback experiments have been a common way of

discovering the meanings of animal sounds, but it is rather a crude and sometimes a disruptive method of research. Whale researcher Peter Tyack once recorded the social calls of male humpback whales competing for the attentions of a female. The sounds seemed to act as a focus for all the males in the area. To check that this was indeed the case, Tyack borrowed an enormous underwater loudspeaker from the US Navy and blasted the calls across the ocean. He was in for a surprise. The whales stopped singing and charged his small boat at high speed. Reaching underwater speeds of 12km/h, the whales submerged and dived just five metres from the boat, and then circled for some time, 'looking for the action'. Tyack does not know what he had 'said' via his underwater playback, but it certainly got a response.

A classic playback experiment was carried out in 1975 by workers at Oxford University. Population ecologist, John Krebs, wanted to know whether the songs sung by great tits in an Oxfordshire wood were territorial 'keep out' signals. In a preliminary experiment he had taken birds from their territories and noticed that other birds quickly arrived from the immediate neighbourhood to take their place. Was it the absence of their songs that indicated the territory had been vacated?

His experiment was simple. He took away some birds as before, but replaced them with loudspeakers. On some he played back great tit songs, on others he had recordings of a tin whistle (which had notes at roughly the same frequency as those in the great tit songs); and in the rest he played nothing at all. Then he sat and watched.

The territories with the great tit recordings were avoided by prospective interlopers, but those with the tin whistle or silence were visited by other males, who took up occupation. Krebs had shown that male great tits waiting to enter and take over territories know whether a territory is occupied or not simply by listening for the resident male's song as a proclamation of ownership. Being an ecologist rather than a animal communication researcher, he did not try to understand all the information contained in the song, but he *had* associated the song with at least one function. By playing an animal's message back to itself, Krebs had begun to unravel what was in the message.

Haven Wiley and Douglas Richards, at the University of North Carolina, used playback experiments to unscramble parts of a bird's song. They were convinced that birds have an 'alerting' component in the song – a simple message that said to other birds of the same species: 'Stand-by, I have a message for you.' A bird has to feed, to watch out for predators, and to attend to a mate or offspring. It does not have time to hang about waiting for another bird to send messages. It must be alert to potentially useful messages in the general background noise of its environment, yet not be preoccupied with communication. A receiver could operate with a low level of vigilance if any interesting message was preceded by an alerting signal. This would have to be an easily detectable and recognizable part of the song or call, which sounded clearly through the forest or other habitat, and had high contrast with any background noise.

Once a receiver's attention was gained by the alerting component, it could switch its attention from whatever it had been doing previously to a short 'time-window' in which was contained the main, probably more complicated, message – perhaps information about the species or the individual making the call, and the state, emotional or otherwise, that it was in. Wiley and Richards thought that many bird songs were designed this way and Richards devised a clever experiment to test their hypothesis.

The subject on this occasion was the rufous-sided towhee, a North American finch which begins its song with one or two simple whistles and then follows with a complex trill. Towhee song played back within a resident's territory will evoke counter-singing. Richards played back different sections of towhee song, including the suspected alerting component only, the message component only, as well as the complete song; some of the recordings were degraded to simulate noisy and reverberent forest conditions. The results were interesting.

If the degraded opening whistle or the degraded trill was played to the resident bird, there would be little response. A clean trill, without reverberation, evoked a strong response. The key experiment came with playbacks of clean and degraded full song. A clean version of the entire song, not unexpectedly, received a strong response, but what was exciting was that so, too, did a degraded version. Wiley and Richards were able to conclude that an alerting component in towhee song *does* allow another towhee to recognize that the signal is coming from a rival towhee, and permits the receiver to get the message in the face of high levels of degradation and noise.

Other researchers have used the same playback techniques to unravel even more detailed information contained in calls. The tiny mud-puddle frog lives in shallow ponds on Barro Colorado Island, Panama. As part of the courtship and mating procedure male frogs call, but not always with the same call. A solitary frog calls 'aow-aow' about 7000 times in an evening. If it is with other males, it calls 'aow-chuck' or 'aow-chuck-chuck', with sometimes as many as five or six 'chucks' in a row.

Stanley Rand of the Smithsonian Institute, Washington DC, was intrigued to know why the frogs used the two different calls, so he carried out a series of play-back experiments. He recorded both 'aow' and 'aow-chuck-chuck' calls and played them back to females. They moved towards the loudspeakers when both calls were played, although they seemed to prefer the latter.

Then Rand edited out the 'aow' sound, leaving just 'chuck-chuck'. The females were not interested. The 'chuck-chuck' calls are an enhancement, concluded Rand, that the male frogs use to gain some advantage over their rivals. The hard-edged 'chuck-chuck' is easier to locate, making the calling male easier to find. But that was not all.

Also working at Barro Colorado was Michael Ryan. He

worked up a series of 'doctored' playback tapes and, while keeping the pitch of the 'aow' call the same, raised and lowered the pitch of the 'chuck' sound. Given a choice, the females went for the deeper chucks, which were most likely a vocal indication of a larger and fitter frog with which to mate. But in the wild there is a price to pay for being easy to find, and the frogs' code was broken by a frog-eating bat.

Using playback experiments, Merlin Tuttle, of the Milwaukee Museum, discovered that the fringe-lipped bat homed-in on the calls of the mud-puddle frog. In the tests they even savagely attacked the loudspeakers, ripping at the covering in an attempt to reach the frog they thought was inside.

It is unlikely that mud-puddle frogs have anything profound to say; nevertheless, the experimenters established that these frogs have a small but significant repertoire of sounds, and they were able to work out what function those sounds performed.

Size and strength featured in the calls of a larger creature studied by researchers from Cambridge University. They used playback experiments to ascertain what red deer stags were 'saying' to each other during the rut on the isle of Rhum, off the west coast of Scotland. The stag's head is adorned with an array of antlers that could inflict serious injury on a rival. There would be a clear survival advantage to a red deer population if they could avoid having to use this formidable weapon. What is required is a reliable and yet safe method of assessing rivals, some sort of contest which would enable individuals to decide who would be the winner of a fight – but without staging the fight itself.

Stags roar up to three times a minute during the rut. In the course of the three-week rutting season they do not feed, but spend the entire period mating and defending their harem against the other stags. Their body weight declines gradually, until they are too weak to rut. Tim Clutton-Brock, the leader of the Cambridge team, was convinced that, with this scale of energy-depletion involved, the roaring contests may serve a critical function.

In order to understand roaring, the research teams looked at the circumstances in which stags roared. When two harems get close together, the harem holder of one approaches the harem holder of the other. One stag roars, the other replies. The two may then roar at each other for over an hour, the roaring rate gradually increasing. Usually, one stag eventually withdraws. If the challenge continues the stags begin parallel walking, a kind of second stage of assessment. If this fails to rectify the situation and one of the stags does not stand down, a fight starts. But it does not often reach that stage. The roaring usually sorts out which is the tougher. Clutton-Brock wondered whether the contestants roared to see which could roar the most or merely because they felt aggressive. Or were they trading roar for roar in a finely tuned system for assessing physical strength?

The researchers taped the roaring of stags for playback experiments. When played to a stag the animal would face the speaker and roar back at it, at the same time increasing its rate of roaring. So far, so good. What the research team wanted to know next was whether the stags increased their rate of their own roaring to match that of their challengers.

In the next series of playback experiments the tapes were edited to give three rates of roaring, each rate gradually increased to give a graded series. The first series was at a relatively slow rate – two and a half roars per minute (RPM). Then a rest period was introduced, followed by roars at five RPM; finally, after another five-minute rest, a tape was played with calls at 10 RPM. In the experiment, six stags were presented with the playback roars. They easily out-roared the two and a half RPM playback, worked a little harder on the five RPM, but at the 10 RPM they tended to stop roaring and in some cases they even rounded up the harem and made a rapid exit. And information gained from field observations of fighting stags by Fiona Guinness and Robert Gibson during the previous five years confirmed that the stags with the highest roaring rate were, indeed, the best fighters. The rate at which a stag roars, therefore, should give an accurate and reliable assessment of how the animal might perform in a fight, and in most cases that information contained in the roars was enough to prevent such a fight taking place.

In experiments with small New Word monkeys, the Smithsonian Institution's John Robinson used playback experiments in order to find out what these little creatures were saying to each other. The results he obtained not only revealed the messages passing between individuals but also reflected their different lifestyles.

One of his subjects was the dusky titi, a primate with a song like a bird. Pairs of dusky titis live in well-established territories and each morning the resident pair visit the boundaries of their home and declare their occupancy with a duet so closely sung it is difficult to hear that there are two animals singing. In simply playback experiments, Robinson showed that duets played at the edge of a territory evoked a noisy response from the residents. Interestingly, female dusky titis call at their female rivals and males sing at males. If he played female calls only or a duet led by the female, for example, the resident female would approach the playback speaker, stare at it, and sometimes pick it up and shake it violently. If male calls were played the male responded, but not so strongly. The female, it seems, has more at stake in keeping a territory and raising young.

On one occasion Robinson decided to confuse a pair of his research animals. He played back the calls of a rival couple, but instead of placing the speaker at the edge of the territory, he put it in the middle. The residents were dumbfounded. The sudden appearance of rivals at the centre of the territory was quite beyond their experience and they failed to respond at all.

Another species, the widow titi, responded in a totally different way to playback experiments. When one was played a rival's call, it simply cleared out of the area. The widow titi is not territorial like its dusky cousin. Its song is used not to defend a territory but to space out individuals throughout a large home range. Instead of the call saying 'stay clear, this area is occupied', like that of the dusky titi, the widow titi's call says 'stay away from me and I will stay away from you'.

The two behaviour patterns can be explained in terms of economics. Dusky titis live in food-rich woodlands; widow titis live on the almost sterile white sands of the Amazon, where there is little nutrient in the trees. Widow titis, unlike their richer relatives, must range very widely in order to find something to eat. They would be unable to defend such an enormous territory with a song. Their songs keep them spaced away from their neighbours.

So, despite the crudity of playback experiments, they help us understand the effect which a particular call might have on an individual. It is one way in which we begin to see the links between calling and the behaviour with which it is associated.

The Dance of the Bees

Of all the non-human animals, scientists have considered the humble honey bee – an animal with no more than a few thousand neurones in its 'brain' – to be one of the few that might aspire to possessing a 'language'.

In 1967 the Austrian zoologist Karl von Frisch introduced the scientific community to the way in which foraging honey bees tell their hive mates of the whereabouts of food. The messages are passed between individuals in the form of a 'dance', and during this dance a bee gives information about the direction in which a rich source of food may be found with respect to environmental cues, mainly from the position of the sun.

During the spring and summer, on days suitable for foraging, worker bees scour the countryside surrounding the hive in search of flowers from which they obtain the pollen and nectar that provides energy for the members of the colony and food to raise new individuals. As a foraging worker returns to the hive, she is first 'frisked' at the entrance to ensure she is a genuine colony member. Only those bees with the correct 'smell' – an important communication channel for bees – are allowed in; others are rejected or even killed.

When inside the hive, the worker then begins to behave in a strange way. She runs about attracting the attention of some of the other workers nearby. They crowd around her, pommelling her with their antennae; and, having attracted an audience, she begins her dance.

If the food she has found is close to the hive, the dance is simple. She performs the 'round dance', moving in a circular pattern first one way and then the other. The number of reversals of directions and the vigour with which the dance is performed indicates the richness of the food source. It does not, however, give any indication of the food's exact location. Presumably, if it is nearby, the other workers can chance upon it easily.

Distant food sources require a different approach. The returning worker begins her gyrations on the honeycomb but this time in a more complex pattern. She moves across the comb at a particular angle, waggling her abdomen as she goes. At the end of the straight part of the dance, she stops waggling and circles to the right, back to her starting point. The dance, known appropriately as the 'waggle dance', is repeated, but this time she circles to the left. The important part of the dance is the straight waggle. The angle that the waggle part makes to the vertical represents the angle between the sun's azimuth and the food source. The length of the waggle section and the number of waggles shows the distance, and the intensity with which the waggles are performed indicates the richness of the source. The waggle dance is accompanied by a buzzing that is picked up by the other workers' legs and antennae. This is also thought to present information about distance and quality.

The entire performance is all the more remarkable when

you consider that the bee must convert horizontal information into gravity-related vertical movements, for the honeycombs, on which she performs her dance, are set upright in the hive. A food source found directly below the sun will elicit an upward-moving vertical waggle dance. Food at 45° to the east of the position of the sun will result in a dance that is 45° to the left of the vertical and, similarly, one that is 45° to the west will result in a dance that is 45° right of the vertical. A food source in the opposite direction to the sun stimulates a dance in which the bee runs directly down the comb.

But that is not all. Research has shown that the bearings flown by the bee to and from the food source do not exactly match the angle of the dance. The angle of the dance is disrupted slightly and offset by the lines of the earth's magnetic field. Bees, it is thought, can detect and appreciate the force of magnetism.

In the body of the bee, bands of cells containing magnetite associated with nerve fibres have been found concentrated around the nerve centre (ganglion) in each abdominal segment. From each ganglion a nerve branch enters the magnetic tissue. It is likely that the tiny magnetic particles twist in response to the earth's magnetic field and produce a torque which can be detected by the accompanying nerve cells.

What happens when the sun goes in? The answer is that bees have back-up systems. On cloudy days, with patches of blue sky, or in forests where the sun might be hidden behind foliage, foragers analyse the polarization of ultraviolet light from the clear parts of the sky and can work out a bearing on the position of the sun which they can then use in the dance. On overcast or foggy days, bees still forage and dance. They are able to predict the position of the sun with reference to local landmarks. And, if there are no useful landmarks, they can rely on the earth's magnetic field alone.

One other remarkable achievement is that the bees can be persuaded, by artificially reducing food availability, to dance in the night – something they would rarely do. A bee, for instance, may be dancing at the end of the day, passing on the desperately needed information to the rest of the workers. As night is falling, the bees will not be in a position to exploit the source immediately and so must wait until morning. The fascinating thing is that when morning comes, the bees head off in the correct direction even though the sun is in a different place in the sky than it was when the forager danced the previous evening – an astonishing piece of 'computation'.

Passing on information about food is not the only conversations that bees have when they dance. They also show this behaviour when house-hunting. After a successful period of rearing young, the colony becomes overcrowded. A new queen is reared and the old queen, with her entourage of old and faithful subjects, leaves the colony in order to set up a new one. It is called swarming.

At first, the swarm does not fly far. It gathers in a great seething mass, sometimes containing as many as 15,000 individuals, hanging from a convenient spot, such as the branch of a tree. Here the bees wait and rest while scout workers search for a suitable place to set up home. This reconnaissance party usually includes the older, wiser workers which know the area well from their recent foraging expeditions.

When one has found a site, it returns to the great ball of bees and performs a zigzag dance, buzzing as it goes. Again, the line of the zigzag indicates the direction and distance of the potential new premises and the tempo of the buzzing gives an indication of its suitability.

Bees are very fussy about where they live. They need a cavity large enough to build sufficient honeycombs for the winter, but not so large that it gets too cold. The entrance must be small and defensible and at least two metres above the ground. It must also face south in order that foragers can warm-up quickly on cold days. An entrance hole at the bottom, rather than at the top, is preferable to reduce heat loss. A scout will take about 40 minutes to inspect a site and weigh up all its pros and cons.

Then she flies back to the swarm with the good news. But, most likely she is not alone; many other scouts will be returning, also with suitable finds. At this point, the bees go into conference. Some scouts dance more vigorously than others. It is as if they have more confidence in the site they have found than their colleagues. If a restrained dancer meets an enthusiastic one, the former races off to inspect the latter's site. Eventually, all the scouts will have inspected the site of the most enthusiastic dancer and, having reached a consensus, all the scouts will be dancing the same dance.

While this has been happening the rest of the swarm simply shuts down and rests. But, when the scouts reach agreement, they move off, one layer at a time, and follow the scouts to the new site. The scouts fly across the slowly moving swarm, indicating the direction in which they all should head. At the new site, the swarm circles around and the scouts drop down. They produce pheromones (chemical messengers) that attract the rest of the swarm, which gathers around quickly until, at an invisible signal, they occupy the new site, clear out debris, and begin to build their combs. They have a new home.

The intriguing thing about the whole swarming exercise is not just the dance and the directional and qualitative information contained in it, but the fact that the scout bees both 'speak' and 'listen'. Clever though it may seem, however, all the bee is doing is putting into action an automatic, innate system of actions and responses by which it obtains and passes on information. When a forager bee arrives at a flower, for example, she observes colour, shape, smell, location, nearby landmarks, when the nectar is produced, the best way to get to the flower and enter the flower-head, and so on. Not every component, however, is learned with the same degree of confidence. Experiments have shown that odour, for instance, is learned at the first visit, whereas colour and location are learned precisely only after several visits. The system is so programmed that if the smell, say, is changed experimentally, the bee must go back and relearn the entire sequence. There is little flexibility. The bee does not seem to be aware of what it is doing, it just does it. But there seem to be exceptions to this automatic, pre-programmed behaviour.

Writing in *New Scientist* in April 1983, James and Carol Gould drew attention to several occasions when bees seemed to be 'thinking'. Alfalfa flowers, they observed, possess spring-loaded anthers. When an insect enters the flower, it gets bashed on the head as the spring mechanism releases. Big, tough, furry bumble bees merely shrug off the blow. Honey bees, on the other hand, seem not to like the prospect of getting hit every time they enter an alfalfa flower; and if released in an alfalfa field they will fly vast distances to find alternative sources of nectar and pollen. Only when desperate will they enter an alfalfa flower, and only then will they pick on the flowers whose anthers have already been sprung. Some have even been observed to chew a hole at the back of the flower and, by using a backdoor, to avoid the spring.

At first sight it seems as if the bees are making a conscious decision to avoid the flowers; but the Goulds point out that the bees' instinctive repertoire could include automatic back-up systems similar to those that come into play in the bee dance when the sun is obscured.

Another example is more difficult to explain in terms purely of instinct. In experiments, sources of food which the bees were encouraged to visit were moved gradually further and further away from the hive. The bees, therefore, had to fly gradually further away. Imagine the surprise of the experimenters when they found that some bees had *anticipated* the next move and were sitting there waiting for the food source to arrive! There is nothing in the bees' normal lifestyle, the Goulds point out, that would pre-programme the creature for such behaviour.

The Goulds reported on another intriguing experiment, this one carried out in 1980. They trained some bees to forage along a lakeside and then tricked the returning bees to dance co-ordinates that placed the food source in the middle of the lake. The new recruits learned the location from the first forager – but they refused to search for the food. In other words, their sense of distance and direction told them that the location could not possibly be a source of food. When the location was changed to the far-side of the lake, the bees began to forage again. They evidently knew how wide the lake was. They each must have had quite a detailed mental map of their environment and they knew that flowers do not usually grow in the middle of lakes.

In another experiment, foraging bees on their way to a natural food source were intercepted and taken to an artificial feeder in the middle of a car park some distance from the hive. The majority of bees, mostly older and more experienced scouts, fed and then flew directly back to the hive even though it was out of sight of the feeder. Young bees simply got lost. The interesting thing here is that the older bees had seemed to recognize the car park – a barren desert of a place – and had it marked on their mental map. They even danced the co-ordinates when they returned to indicate to other bees that there was food in this normally sterile environment.

Does an animal like a bee, which has its behaviour patterns hard-wired (to use computer jargon) for the most part, have the capacity to fall back on what we call 'thinking' when a system fails to cope? And might a bee's dance 'language' be a component in that process of thinking? In other words, is this an example of language leading to thought? It seems that although most of a bee's behaviour is innately programmed and intellectually undemanding, there are a few examples to show that, like ourselves, 'a bee is just as clever as it needs to be'.

Cheating

Though not as complex as the bee dance, the behaviour of the *femme fatale* firefly is one which illustrates how, in an evolutionary race, mastery of a channel of communication can gain one organism advantage over another. In this case the medium is not sound but vision. Fireflies have a sign 'language' which, like morse code, is a system of flashes. The flashes are used to bring males and females together. The message in the signal is simple – 'I am species X, and I am ready to mate'.

Over 3000 years ago Chinese poets drew their readers' attention to the light of the firefly. Much later, Aristotle recognized that there were certain creatures that 'give light in the dark'. Shakespeare was not so impressed, referring to the glow-worm's 'ineffectual fire'. The Spanish conquistadors, however, discovered that the peoples of the West Indies obtained their illumination from that very fire. The click beetle *Pyrophorus* provided the light, whether in the home or as adornment in the hair or on the clothes.

Although given the name glow-*worm* or fire-*fly*, the organism that provides the light is neither a worm nor a fly but a beetle. The luminosity it produces is an aid to courtship. The female European glow-worm *Lampyris* does not flash. It is flightless, looks more like a beetle larva than an adult, glows in the dark, and thus attracts the male towards her. North American fireflies, however, flash at each other. Each species has a different pattern of flashes that cannot be confused with that of another species.

Male *Photinus pyralis* fireflies, for example, fly a zig-zag course about 50cm above the ground, drop down every 5.8 seconds and give a flash that lasts for about half a second. Any female within two metres waits two seconds and flashes back. *Photinus consimilis* males are slow flashers. Flashes are delivered in groups of three while the insect is about three to four metres above the ground. *Photinus granulatus* swoops close to the ground and bounces from side to side, while *Photinus consanguineus* makes two short flashes two seconds apart, which are repeated every five seconds. Some species, however, take advantage of the flashing game by cheating.

At the University of Florida at Gainsville, James Lloyd and Steven Wing swung light-emitting diodes and found that they attracted female *Photuris* spp. fireflies. But these ladies were not interested in mating, they actually attacked the light sources, homing-in like heat-seeking missiles. Lloyd and Wing looked more closely at the behaviour and found that not only did they flash the correct code for males of their own species, but also they mimicked the code of a related species of *Photinus*. But when the *Photinus* males descended they were in for a deadly surprise. They did not find the female of their choice but a firefly *femme fatale* that simply gobbled them up. The *Photuris* females had cheated the *Photinus* males not only to gain extra nutrients during egg-laying, but also to obtain noxious predator-repulsing chemicals that the *Photinus* males contained. They thus gained some measure of protection against wolf spiders and other predators.

A Stranger in the Camp

We have seen that related species of animals can communicate with each other. But the idea of a butterfly 'talking' to ants seems to belong to the realm of fantasy. Yet for the larval, pupal and adult phases of the large blue butterfly it is not fantasy but a stark question of survival: they must 'talk' their way out of danger and convince their kidnappers that they really are worth more to keep than to kill.

The revelations about ant and butterfly talk were made by Jeremy Thomas at the Institute for Terrestrial Ecology at Furzebrook, Dorset. He had been following up the work of the English naturalists Bagwell Purefroy and F W Frohawk who, in 1915, had noticed that the large blue butterfly spent the first part of its caterpillar stage as a vegetarian feeding on thyme plants, but in the later part it changed to a carnivorous way of life and sought the company of red ants.

In the autumn, the caterpillar encourages red ants to carry it back to their nest. It does this by rearing up and producing a blob of a honey-like secretion from one end. The ants find this irresistible and carry their find back to the nest, where they palpate the caterpillar's skin with their antennae in order to encourage it to produce more of the sugary solution. The caterpillar 'talks' to the ants in their own 'smell language', producing pheromones (chemical messages) that tell the ants 'I'm OK, don't harm me'. It is an appeasement signal that suppresses the ants' natural tendency to tear the caterpillar to pieces. Unfortunately for the ants, the caterpillar is lying: it repays its hosts' hospitality by gobbling up their larvae.

After six or seven weeks, with the onset of winter, the caterpillar changes into a pupa. It will spend the winter in the comfort and safety of the nest, protected by the ants. In order to keep the ants happy it must continue to talk.

Ants talk to one another through a variety of channels of communication – smell, taste, vibrations and sound. The sounds, for example, are made when they rub their back legs together. The large blue butterfly, amazingly, has cracked the code, and from inside the chrysalis the metamorphosing butterfly copies the ant sounds by tapping on the pupal case. In short, not only has the butterfly tuned-in and copied the ants' smell and taste signals, it has also worked out the subtleties of their complex sound system. And there is more.

As spring approaches and it is time for the chrysalis to break open and release the adult, the developing butterfly increases the rate of tapping, causing the ants to gather round. Having assembled its bodyguards, the butterfly emerges. It is not attacked, for once again it produces appeasement secretions that pacify the ants. It then heads for the surface with its entourage following close behind. An adult butterfly is vulnerable to passing predators, such as ground beetles. When it emerges, it must climb and wait at the top of a grass stem for its wings to dry before it can fly safely away. The ants, encouraged by sugars and smells, provide an armed escort for the vulnerable butterfly. This is a remarkable example of interspecies communication.

Conversing in Living Colour

Imagine being able to soothe a potential mate, frighten-off a predator, warn a rival, and say that you are frightened – all at the same time. Squids can do that, not with words but with body talk.

Squids, cuttlefish and octopuses are among the great conversationalists of the deep. By changing the designs, patterns and colours on their body they can hold many conversations at the same time. Indeed, this has led Martin Moynihan and Arcadio Rodaniche, at the Smithsonian Tropical Research Institute in Panama, to consider that squids and cuttlefishes might have a larger vocabulary with which to speak than we do.

In human language there is a limited number of vocal sounds or phonemes that we can use, and these in turn can be integrated in a limited number of ways. With the subtlety of colour changes that a squid or cuttlefish can achieve, coupled with the fact that several colour patterns and changes can be combined simultaneously – something that we cannot do with our phonemes – it is quite possible that these seemingly primitive creatures have a 'language' that rivals that of humans.

Cephalopods, which include the squid, the cuttlefish, the octopus, the pearly nautilus and the argonaut, are in fact not primitive in the least. They are at the top of the invertebrate tree and are considered to be the most 'intelligent' of the animals without backbones. The cephalopod brain is large. Compared to body size, it occupies a place equivalent in the vertebrate hierarchy to that of a reptile, although some consider it to be a great deal more clever. Its primary sense is vision – the cephalopod eye is not unlike our own, with the capability of appreciating colour.

Recognition of the abilities of squids and octopuses goes back at least to classical times. In the Mediterranean they were not only considered a culinary delicacy, but were also recognized for their cunning. Aristotle was of the opinion that they were 'more than a match for fishes even of the larger species'. The early scientists were also aware of the way that certain of the cephalopods changed colour when angry or frightened: they seemed to wear their emotions on their sleeve, as it were.

In the skin of a cephalopod there are thousands of elastic-walled colour cells which can be filled or drained of pigment within seconds. In fact, it is probable that the cephalopods are capable of the fastest changes of colour in the animal kingdom. This is done with the aid of a complex series of nerve endings and tiny muscle fibres that surround each cell and which are directly connected to the animal's brain. Each cell can be contracted individually, so there can be a conscious choice of which cells to make dark and which to keep light. Any little thought results in a colour change, and it might show as a small patch on one part of the body or as a complete change of body colour.

In the BBC television film *Aliens of Inner Space* cuttlefish and squid showed the most extraordinary colour changes. A Texas Researcher, Robert Hanlon, introduced viewers to dramatic transformations of skin colour: longitudinal or diagonal 'zebra' stripes, dark transverse bars, spots and blotches, moving patterns of colour that started at the animal's rear and travelled the length of its body. Hanlon felt that these patterns give us some insights into the workings of the cephalopod brain:

> The skin of the cephalopods is a perfect two-dimensional view of the brain. What you have are nerves coming out of the chromatophore lobes, and when innovation takes place, you get something on the skin that says there is nervous activity. So, in fact, a pattern is a window into the brain.

Hanlon has shown that these colour cells, or chromatophores, are not present in the early squid embryo. The small spots of living jewellery begin to appear only when the nervous system starts to develop. At first they have only a covering of polka dots but gradually the colour patterns become more elaborate as the squid gets older.

> The truth is we're only beginning to understand the range and complexity of the patterns, and we are even further yet from understanding what they mean and how they are used. The patterns are simply too complicated to describe ... they change rapidly, they grade one into the other, and it is too complex for the human mind to comprehend what one pattern is.

In fact, the combinations of colours in the patterns are infinite. In a simple scenario an animal may have three colours which can be displayed with different grades of expansion of the colour cells. And they can be overlapping.

Some of the patterns can be interpreted. The wave of colour that passes from the back to the front can mesmerize prey or confuse a predator. When a cuttlefish approaches a shrimp, for example, it will face the prey, raise its arms, and then show the moving wave pattern. The shrimp is frozen to the spot, for it sees something that it has never experienced before, and in that moment, the long tentacles of the cuttlefish shoot out and grab it.

There is, however, a belief that, contained in these visual displays, are more complex messages. And what Moynihan and Rodaniche began to realize was that these extraordinary patterns could conceivably be the component parts of a cephalopod 'language', with its own syntax and grammar.

They had been watching the Caribbean reef squid and identified at least 35 distinct patterns. Pale patterns seem to relate to fright and the need to escape. Dark patterns are associated with aggressive behaviour. Soft 'pastels' appear to have sexual connotations; they are so subtle that they can be seen only when two animals are close together. Stripes, on the other hand, are startling, mainly to predators, and it is significant that many of the distinctive displays identified by Moynihan and Rodaniche serve this function. Squids have many enemies, and so have adopted what Moynihan calls 'anti-displays' with which to confuse predators.

Moynihan's previous work was with bird and primate communication and, together with Rodaniche and others, he has been trying to discover whether the cephalopod visual communication system has patterns or colours which we might consider to be the equivalent of verbs, nouns and adjectives or modifiers. To this end, he has been using the same computer programmes as those designed to analyse language.

The researchers have fed in hundreds of hours of observations in order to relate 'words' (the colour changes) with 'actions' (the behaviour of the cephalopod or other species nearby). So far, they have recognized three categories of signals: signifiers, modifiers and positionals. The signifiers are the nouns and verbs and are the strongest messages. They are usually associated with things like attack or escape, or are linked to food. Modifiers, such as belly stripes or fin stripes, contain less information and always accompany other signals as if they qualify or reinforce them. Positionals, such as head-down displays or V-contortions, in which the animal's body or parts of the body are used, seem to be the weakest signals. They appear to provide the stage on which the colour changes take place, rather than qualify the message.

The patterns appear to be not random but selected to accompany particular pieces of behaviour. In his book *Communication and Noncommunication by Cephalopods*, Moynihan tells of one intriguing case. A group of young reef squid, when approached by a predator, all contort into the V-posture and show dark stripes. Or, rather, all do this except for the one nearest the predator: he, on the contrary, displays a quite different pattern. As Moynihan writes: 'If last, go odd; if odd, go last.' By being different, the most vulnerable one in the group confuses the predator. The predator, probably programmed to react to the escape behaviour (V-posture, dark stripes) it has learned to associate with this species, is expecting one pattern but is presented with another. For an invertebrate which is distantly related to slugs and snails, that seems quite a clever thing to do.

A Fishy Tale

Fish and shellfish are great talkers: cod and haddock serenade their intended with a series of grunts that rapidly increase in tempo as things hot-up; pistol shrimps fill the oceans with a staccato chorus of loud and incessant clicks; lobsters rasp away in the confines of their rocky home; grunt fish, not surprisingly, 'grunt'; trigger, drum and croaker fish rasp their swimming bladders like a finger drawn over the surface of a balloon; and sea-horses and pipe fish rub the back of the skull against a projection on the top vertebra to make a clicking sound. The messages contained in such signals are relatively simple: they are either to warn a rival or to attract a mate.

Little was known about the love-torn haddock until Anthony Hawkins, at Aberdeen's Marine Laboratory of the Department of Agriculture and Fisheries for Scotland, watched a pair in a 700-gallon aquarium. The haddock, it was revealed, is quite a conversationalist and woos his intended with an elaborate display of vocal and visual signals.

The sounds are single repeated knocks and pulses of up to 16 grunts. Knocks seem to show the fish's level of excitement and are given in aggressive situations. Grunts are heard after the confrontation is over. In the spawning season, male haddock call continuously, some calls seeing off rivals and others attracting females. A hydrophone dropped into the North Sea during February picks up a cacophony of haddock love-calls.

When two rivals sidle up to one another the knock calls are delivered first at a slow rate. As the antagonism between them increases the sounds are delivered so fast that they become a continuous hum. The males first face and then swim parallel to each other, their fins spread out in display. Dominant fish ram subordinates and chase them away. A male that has fought his way to become top fish swims in a tight circle around the female, his knocking sounds becoming faster and faster until the sound is like a continuous purring, rather like a motorbike starting up and moving off. At the same time a marked change occurs in his appearance. Dark blotches appear on his sides and the fins become darker in colour. At the moment of spawning the male becomes silent and the pair embrace as they swim upwards. During the nuptial swim, the female lays about 10,000 eggs and the male sheds his milt to fertilize them. Sound, it is believed, is the stimulus for the female to spawn.

Some fish have a novel form of communication: electricity. If our senses could detect the electric and electromagnetic events that take place in the sea and in rivers and lakes, we would experience a whole new world. It is a phenomenon readily available for underwater creatures to use. Some fish can produce their own electricity and send their own electrical messages.

In 1958 Walter Lissman, of Cambridge University, noticed that some fish could produce weak electric fields with which they found their way about their underwater environment. Further research revealed that there were those that could not only electro-locate but also electro-communicate. It is, after all, an effective system for a fish to use: it can be turned on and off instantaneously, can pass through turbid water, and can travel around obstacles such as rocks and submerged roots. It is also surprisingly 'private', for the sense electro-reception is not common among sea creatures; there will be few messages jamming the communication channel. There is, however, the problem of distance. The tube-snouted mormyrid fishes from Africa, for instance, have an effective range of 100cm for electro-communication.

The electric current is generated by modified muscles or

nerves. An electric organ may consist of several strips of flattened cells, known as electrocytes, which may occur in various parts of the fish's body. Most of the South American and African weak-electric fishes have modified axial and tail muscles adapted to produce electricity, although one group of neotropical fishes, the apteronotids, have modified nerves. One species of gymnotid, a South American knife-fish, has the anterior and posterior faces of its electrocytes discharging at slightly different times, giving it a species-specific signal. Another knife-fish, which has electrocytes derived from nerve tissue, discharges at the phenomenal rate of 1800 pulses per second – almost the limit for electrical signalling.

Electro-communication has been found to parallel other modes of communication to the extent that 'electro-languages' have developed. Fundamental to communication is the ability to transmit information about the identity of the communicator – its species, sex, stage of development and age. Also useful is motivational information such as readiness to mate, proclamation of territory, spacing signals and information about the quality of a prospective mate. Weak-electric fishes, it turns out, have distinct signal patterns for many of these functions.

There are two types of signallers – the pulse species, with bursts of electrical activity at low discharge, and the wave species, with more continuous, high-frequency discharges that broadcast in species-specific wavebands. Many of the pulsed species transmit their signals in distinct patterns that are repeated over long periods, sometimes as much as 10-20 minutes. The elephant-trunk fish of Africa has a characteristically coded discharge rate which rises slightly after 25 milliseconds, then drops and rises again steeply at 100 milliseconds, and then tails off at 250 milliseconds.

The knife-fish are wave species. The muscle-derived signals are broadcast at 50-150Hz, while the nerve-derived ones broadcast at 750-1250Hz. Within that range, males and females have different frequencies. *Sternopygus* spp. have muscle-derived electrocytes, and the males and females signal on different frequencies – males on 50-90Hz, females on 100-150Hz. Youngsters living in the nest broadcast on yet another frequency. The frequencies rise during sexual or aggressive encounters. If two fishes broadcasting on the same frequency meet, they avoid jamming each other's signals by altering their frequencies slightly.

The more we study animal communication systems the more surprises we uncover. Marine researchers have tuned into at least one example of 'shark-speak'. It is not 'spoken' sound, smell or electricity but is a part of shark 'body language'. The message was delivered by a Pacific grey reef shark to a US shark researcher, Don Nelson, who was approaching it in a small submersible. The shark, evidently thinking that the craft was too close, gave a warning to stay clear. The threat display was in the form of an exaggerated swimming pattern. The shark's nose went up, its pectoral fins turned down and the body was contorted into an S-shape. When the signal was ignored, it followed with a head-on, slashing attack with its mouth wide open. That kind of body language one ignores at one's peril!

Singing Whales

Irish legend has it that in the sixth-century St Brendan was sailing in search of 'that paradise amid the waves of the sea' when he came upon a mysterious island around which he could hear three choirs singing hymns. Similarly in the eighth century Maeldune was searching the oceans for his father's murderers and chanced upon an island from which he could hear the singing of psalms. And when Usheen the legendary folk hero reached the Island of Youth (thought to be somewhere to the west of Ireland) he heard what he thought were birds singing.

As in all legends, there is an element of truth in these stories. Irish monks are thought to have made long journeys across the Atlantic, taking them even as far as North

America, and their progress westwards could have taken them close to the island of Bermuda. And in this location to this very day, you can hear the strange and mysterious 'singing' that is most likely the source of all those ancient legends. The animals responsible are not birds, as the storytellers believed, but whales; in particular, humpback whales.

The whales come to Bermuda to breed and it is during the breeding season that they sing their eerie and melancholy songs. They are part of the north-western Atlantic population of humpbacks that spends the summer months feeding in the Arctic and the winter months around Bermuda or further south in the West Indies. Other populations make similar long journeys from feeding to breeding grounds. In the eastern part of the northern Pacific humpbacks feed off the coast of Alaska and breed at either the Baja California peninsula or Hawaii. At the same time in the southern half of the eastern Pacific another distinct population travels in the opposite direction, spending the Austral summer feeding in the Antarctic and the winter mating and calving in the warm waters off the Central American coast.

The early whalers must have heard the whale songs. There are few records in ships' logs or crews' diaries to say that they did, although in their fascinating book *Wings in the Sea*, Lois and Howard Winn note one entry in 1883 in the log of the barque *Gay Head*. The reference is to a 'singer' being caught in the Gulf of Parita on the Pacific coast of Panama. The date was 15 August, the middle of the southern hemisphere winter, when the Antarctic whales are on their breeding grounds in the tropics.

Apart from this single entry, there was no hint that the legendary singers were whales until 1952, when O W Schreiber recorded whale songs from a US Navy underwater listening post on Hawaii. At the time he did not identify the animal responsible for the sounds, but this was done later by William Shevill of the Woods Hole Oceanographic Institution at Falmouth, Massachusetts.

The next important date in the story of humpback whale

song was in 1967, when Roger and Katherine Payne met Frank Watlington, an acoustics engineer at Columbia University's Geophysical Field Station. Watlington recorded anything that made a noise under the sea, but was particularly interested in underwater explosions. Off the coast of Bermuda he had recorded humpback whale sounds and he played them to the Paynes. They were fascinated by the moans and whines and went in search of the creatures that had made them. They rented a sail boat and tried to get close to the whales without frightening them. Then, one day, they heard the sounds of whales being amplified through the bottom of the boat and noticed that some of the phrases sounded similar to the sounds they had heard on the Watlington tapes. They realized that they were not listening to random sounds but to regular repeating patterns: the humpbacks were singing true songs.

Humpback songs, the Paynes revealed, consist of long, complicated, repetitive sequences, much like bird song except that each one can last from five to more than 30 minutes. A whale may start, stop, or resume singing at any point in a song, so it is difficult to determine the beginning, middle and end. Each song is sung over and over again without breaks, except for short breathing spells, for many hours. A whale recorded in the West Indies by Howard and Lois Winn of the University of Rhode Island, sang non-stop for 22 hours! It was still going strong when the Winns pulled up their hydrophones and went home. They are undeniably the longest songs known.

Off Bermuda in those pioneering days in the late 1960s, the Paynes collected long sequences of whale songs by dangling hydrophones from outriggers on each side of their sail boat. They worked day and night, recording the sounds on tape recorder, and when they felt they had enough material they returned to the United States, where Scott McVay of Princeton University helped in the analysis. Together with the Watlington tapes, they had an almost continuous record of Bermuda humpback whale songs over a period of 20 years.

In an attempt to break the songs down into manageable lengths, the Paynes divided each song into identifiable parts. The smallest part is a unit and is the equivalent of a musical note. Units are grouped into small repeating sequences called phrases. Groups of similar phrases are called themes. The Bermudan whales had eight to 10 themes in each song, whereas (it was later found) Hawaii humpbacks have only four or five themes. The song is repeated without pause as a song pattern. By breaking the songs down in this way and comparing songs from different whales in different years, the Paynes were able to make some remarkable discoveries.

All the humpback whales in a particular area, they found, sing the same song. Whales from the North Atlantic, however, sing a different one from those in the northern Pacific. Those sung by populations of whales in the northern hemisphere are dfferent from those sung in the southern hemisphere. In short, distinct populations have their own song, although the laws governing the songs are the same in all populations.

A song, for example, might consist of five themes A-B-C-D-E which always follow in the same order. If a theme, say C, is omitted for some unknown reason, the remaining themes are always sung in the same order A-B-D-E. It is the same as Western music: if pieces of music are written in New York and London, the laws governing the two compositions are the same.

There was one interesting revelation when scientists analysed the songs of the north-eastern Pacific population. They discovered that the whales off Baja California sing the same songs as those around Hawaii. Using song patterns and the photographic identification of individual body, fin and tail markings, it was found that the whales that feed together in Alaskan waters in summer have a choice of California or Hawaii for breeding in the winter. This revelation is of great significance in whale conservation. It was realized that whales might be counted twice – once in each breeding site – and an overestimate made of the number of whales in a population. This could have had disastrous consequences if the stock was hunted commercially and there were only half as many whales as was thought. The study of whale singing helped to plug a conservation loophole.

Another surprising discovery was that the songs are continually changing. All the whales in a population change their songs in the same way so that each individual is up to date with current vogue – a kind of Top of the Cetacean Pops. The changes are progressive and rapid, each component perhaps changing every two months, although many elements of the song may be changing at any one time. Taking a folk-song analogy, in the first month an individual might sing: 'London's burning, London's burning. Fetch the engines, fetch the engines ...', and so on, while two months later one phrase would be modified so the song became: 'London's burning, London's burning, London's burning. Fetch the engines, fetch the engines ...', etc.

Humpback whales are composers, much as people are, except that they do not create new songs, but evolve variations from what they already have by making minor modifications. During a season a song may change components, add extra parts and drop in pitch. A component may undergo rapid changes for several months and then be left alone while other parts of the song are modified. Curiously, newly created phrases are sung more rapidly than older ones. And sometimes a new phrase is created by taking the first and last parts of older phrases and dropping the bit in the middle, much as we shorten 'I would' to 'I'd'.

What advantage a whale may gain from changing its song is unclear. Whether it is a dominant trend setter that introduces change will be difficult to find out. What is clear is that an entire song is renewed totally after about eight years. This has, in fact, allowed the Paynes and their sponsors, the New York Zoological Society, to identify the recordings they made in Bermuda and which were released on the famous commercial disc *Songs of the Humpback*

Whale. Any unauthorized use of the material can easily be spotted because the whales have never sung those amazing high-pitched 'crying' sounds since 1967.

Humpbacks sing only on the breeding grounds and, occasionally, on migration, although they make all sorts of snorts, whines and burps at other times. Only one isolated and misguided individual has been heard to sing on the feeding ground. What happens to the song, therefore, between breeding seasons? Whales, like elephants, seem never to forget. In the new season the song is picked up, usually as it was left the previous season, and remembered by each male in the population. Most of the changes that take place occur during the singing season and not between.

But which whales do the singing – males or females – and what are they saying? From the start of this research it was assumed that singing was associated with behaviour confined to the breeding grounds. William Shevill, writing in *Marine Bioacoustics* in 1964, had noted that:

> The sonorous moans and screams associated with migrations of [humpback whales] past Bermuda and Hawaii may be audible manifestations of more fundamental urges …

And it is the male humpbacks that make them. There is some evidence to suggest that an assembly of singing male humpbacks is a gigantic ocean lek, where males gather on a communal display ground to which females come in order to breed. It is not clear whether the singing is to tell other males to keep their distance, and therefore a 'spacing mechanism', or whether the quality of the sounds is important for females to choose their mate, so that it becomes a 'choose me because I'm the strongest' signal.

One of the workers who carried out the early work on humpback singing behaviour was Peter Tyack (now at the Woods Hole Oceanographic Institution studying dolphin communication). I remember him telling me what happened when he went out to record the whales.

> When you get into the water near singing humpback whales it can be a scary experience. Your lungs resonate with the sound; it is very loud and you feel it throughout your body – you feel it more than hear it – it is very eerie to be sitting 30ft below the water just having your body reverberate with the sound of the whale. Occasionally, when one of us is alone in a boat and has been out there for a few hours, he'll notice the whale next to the boat, 10 or 15 ft away. The whale will often surface, come right up within inches of the boat itself, seeming to look the boat over, often staying for up to half an hour at a time, spy-hopping, lifting its flippers up; it is very strange to have your wild animal come up to you when you don't expect it.

Often Tyack was able to identify a singing whale because, when it surfaced, the sound level would be reduced. Occasionally he noticed the whale would stop singing and then surface with another whale, the two whales moving off together.

In further research, whale watchers noted that during one season in 1979, 13 out of 28 singing whales joined other whales when they stopped singing. Indeed, when a whale stopped singing the team were ready to expect some interesting behaviour. They were able to build up a picture by recording the sequence of events that took place. Such a sequence might be as follows:

A singing whale is almost always alone, separated by several hundred metres from any other whales. It moves slowly, turning this way and that while singing. Occasionally it moves towards other whales, but it avoids other singers. Singers do not appear to perform duets or interact vocally in any way. The message is one way: a reply is not expected. They also do not seem to patrol strict territories, and singers are not found singing at the same 'song-post' on different days. On some occasions a singing whale has been seen to be approached by a non-singing whale. The singer stops singing, swims rapidly away and the new arrival takes his place and starts to sing.

If a cow and calf come into the vicinity of a singer they do not move towards him. Often as not they will actively avoid him. Whales have a breeding cycle in which the cow conceives one year, has a calf the next, and is unreceptive for a further year while suckling the calf. If, however, the calf has been weaned and the cow is receptive to mating, the following might happen:

The singer will stop singing and start to pursue the cow and calf. If he catches up he becomes escort to the cow. The cow, calf and ex-singer then swim along slowly and silently together, the calf keeping close to the cow. Escort whales were once thought to be 'aunties' or 'helpers', but now it is believed they are potential mates for the cow. As they swim they interact with gentle flippering and rolling. Occasionally the cow and escort leave the calf at the surface and dive together, disappearing for up to 15 minutes. They then rejoin the calf for five minutes or so before diving into the depths again. It is thought that this is when mating occurs.

If the cow, calf and escort swim close to another singing whale another series of events may take place. The singer approaches the group, which speeds up and alters course in order to leave him behind. If he reaches the group things begin to hot up. The group accelerates and the new arrival jostles with the primary escort for the right to take his position. Suddenly there is a barrage of snorts, grunts and loud trumpeting calls as one male rams the other, and they thrust their tail flukes at each other whilst aggressively blowing bubbles. These social sounds are very loud. They can be heard at a distance of over nine kilometres, thus advertising the location of the group. Other ex-singers join the fray and the level of activity increases. At the surface whale watchers have seen heads, flippers and flukes being thrown out of the water. The whales, far from being gentle, are very aggressive, each male trying to occupy the position next to the female by placing itself between the cow (the nuclear animal) and the first male (the principal escort). Bouts are rough. Whales have been seen with 'raw, freshly bruised dorsal fins' and 'head nodules that appeared red and bruised'. Up to 15 animals might be vying for the right to mate with the female.

Sometimes the principal escort is displaced, and occasionally he will regain his position. But he will not stay there for long. According to Hal Whithead, the whale researcher who first described the nuclear animal-principal escort structure, a humpback bull will not keep his position for longer than eight or nine hours.

After the rowdy group has been travelling together and brawling for some time, it begins to break up. One by one the unlucky males peel away and occupy a song post, where they resume their appropriately melancholy song.

Scattered Herds of Ocean Wanderers

For many years a mysterious low frequency sound has been heard in the oceans of the world. The sound consists of pulsed 'blips' with a frequency close to 20Hz. With a bandwidth of only 3Hz, the '20Hz signal', as it is known, is almost pure tone. It is also of high intensity, so loud that scientists at first believed it to be non-biological in origin, such as the noise of surf breaking on a beach. But the sounds increase in late afternoon, reaching a peak around midnight, and waves do not have such a diurnal pattern.

Clues to the identity of the sounds began to appear when B Patterson and G R Hamilton were recording sounds off that favourite location for underwater recordists – Bermuda. They heard trains of pulses of the '20Hz signal' which lasted for about 15 minutes, separated by two-and-a-half-minute periods of silence. They thought the rhythm reminiscent of the breathing cycle of a large, slow-swimming whale.

William Shevill and his colleagues at Woods Hole, using a large microphone array, were able to identify the sender of the '20Hz signal'. It was a fin whale, the second largest and the fastest of the giant baleen whales.

Why, wondered researchers, should fin whales make such

a sound? Other cetaceans, such as dolphins and killer whales (as we shall see later), use sounds to keep in touch with other members of their school or pod. Apart from a cow with a calf, fin whales are usually seen swimming alone or in small groups. Why they should need to make a very loud low-frequency sound was a mystery – that is, until Roger Payne and Douglas Webb came up with the idea that the fin whales were not travelling the oceans on their own but were part of a vast, widely scattered herd in which individuals kept in touch with a low-frequency call. The two researchers do not claim that complex and meaningful messages are being sent across the ocean, simply that acoustic signalling might be used by fin whales to locate one another, to aid rendezvous and to maintain the cohesion of the group in spite of the wide dispersal of its individuals. A seemingly lone whale, therefore, might have a 'travelling companion' a couple of hundred kilometres away over the horizon.

Patterson and Hamilton were able to give some substance to the theory. They listened to two fin whales, several kilometres apart, emitting pulsed '20Hz signals'. Using a multiple hydrophone array they were able to track the direction in which the whales were heading. The first whale called for about three hours while heading south. The second whale, about five kilometres to the east, then began to call, and the first whale changed direction towards it. Was it responding to the other's call? Unfortunatly, any work in the ocean depths is expensive and researchers have still to find the definitive answer to this and related questions, but the theory and the early evidence are tantalizing, to say the least. It may be significant that fin whales, unlike humpbacks and grey whales, do not have readily identifiable breeding sites, and they do not appear always to breed in the same areas. The '20Hz signal' could be one mechanism which brings them together.

The one sobering thought is that the fin whale's communication system would have evolved in a much quieter ocean and it is conceivable that man's maritime developments – steamships followed by diesel, passenger liners followed by supertankers and bulk ore carriers – could have seriously impaired communication between whales and upset their lifestyle – a further setback following their wholesale slaughter at the beginning of this century.

Moby Dick and Friends

The sperm whale is a large toothed whale that produces clicks. William Watkins and William Shevill at Woods Hole have been analysing its sounds. Using a multiple hydrophone array, they can accurately locate the position in the ocean of a 'clicking' whale.

Groups of these large whales are heard to make clicking sounds, and each whale produces its own distinctive pattern of clicks, like a morse code. Using this unique identification and location technique, Watkins and Shevill have been able to track individual sperm whales and observe their behaviour. They have found, for example, that sperm whales surfacing within 10 metres or so of each other spread out in an inverted funnel pattern when they dive again, and so are separated by much greater distances as they reach the bottom. Returning to the surface, they emit more clicks and gather together once again in a close-knit group.

Elephants' Secret Messages

At one time, before the price of ivory rocketed and poachers armed with automatic weapons (the legacy of countless African wars and skirmishes) and unscrupulous officials hoarded 'white gold' as insurance against lean times, there were huge herds of elephants roaming the grasslands of East Africa. In Kenya they shared the savannah with the Masai, nomadic pastoralists whose economy is based on cattle. Elephants are understandably wary of people – any people. The Masai noticed that any elephants they disturbed were somehow able to mobilize all the other elephants in the immediate area. At some hidden signal they would all come running to help and face the perceived threat. The Masai could hear no trumpeting or loud noises, yet elephants from a couple of miles away – some engaged in bathing, perhaps, and others feeding or sleeping – were suddenly alerted and they all headed towards the source of danger. The Masai thought that elephants had a mysterious, telepathic 'secret channel' of communication.

It turns out that the Masai were right: elephants really do have a hidden communication channel – hidden, that is, from people. In 1986 scientists at Cornell University in Ithaca, New York, discovered the nature of that secret channel. It was not ESP or anything as exotic or mysterious. Elephants apparently 'grumble' to each other in very-low-frequency sounds – some so low that humans cannot hear them. The grumblings and rumblings are called infrasounds and are generally below 20Hz. Healthy humans, with good hearing, can detect sounds between 20Hz and 2kHz. Above and below that they hear nothing; indeed, below even 30Hz the sounds must be very intense for us to perceive anything. It is in this infrasonic range that elephants talk to each other.

Infrasounds are not unusual in nature. Waves crashing on the shore, earthquakes, volcanic eruptions, and wind blowing across cave-mouths or canyons all produce infrasound, and there is some speculation that certain animals, such as birds, use the distinctive infrasound signatures associated with geographical features for navigation during migration. Pigeons, for example, have been found to be able to hear infrasounds.

The scientists who revealed the elephants' secret were Katherine Payne, who had already unravelled the song of the humpback whale, William Langbauer, who had some experience of talking with dolphins, and Elizabeth Thomas. They were working at Cornell's Laboratory of Ornithology which is guardian of the extensive Library of Natural Sounds, one of the world's largest collections of wildlife sounds.

Apes in conversation: Roger Fouts and seven-year-old Tatu 'talk' using American sign language at Central Washington University. *(Popperfoto)*

Apes as artists: six-year-old chimpanzee Baerbel completes another masterpiece. Her ape art work sold well at a Frankfurt art gallery. *(Popperfoto)*

The pygmy chimpanzee is considered to be man's nearest relative and the brightest of the apes. *(Oxford Scientific Films – J C Stevenson)*

Wild dolphin and young girl meet at Monkey Mia in Western Australia. A school of wild dolphins can be stroked and fed by people. *(Claire Leimbach)*

The cuttlefish wears its emotions on its skin. Rapid changes of colour in pigment cells offer scientists a window on the cephalopod brain. *(Oxford Scientific Films – Carl Roessler)*

African gray parrots not only 'talk' to each other but can communicate with people too. *(Oxford Scientific Films – Frank Schneidermeyer)*

Clever Hans, the horse that could count, featured in a television drama. The experiments to test the horse's mathematical abilities were reconstructed in the programme. *(BBC)*

A returning forager honey bee tells its hive mates about the best sites to find nectar. It performs a complicated dance.
(Oxford Scientific Films – David Thompson)

Above, a blind-folded assistant at Louis Herman's dolphin facility in Hawaii signs instructions to a captive bottlenose dolphin. *(Walter Sullivan,* New York Times)

Left, the pens at the Dolphin Research Center in Florida have such low fences the dolphin's could 'jump ship' and head out to sea; instead, they stay.

Killer whales in captivity have thrilled audiences and dispelled the myth that they are dangerous to people. They have also taken part in man–animal communication experiments. *(Bruce Coleman)*

Vervet monkeys have alarm calls for specific predators. The snake alarm has monkeys standing upright and looking for the source of danger. *(Oxford Scientific Films – Eyal Bartov)*

Katherine Payne told her story in the August 1989 edition of the *National Geographic* magazine. She wrote that she became aware that something strange was happening in 1984 on a visit to Portland's Metro Washington Park Zoo in Oregon. She was watching three Asian elephants and their calves and at the same time experienced a 'throbbing' feeling, yet she could actually hear nothing. It was like the throbbing she had experienced as a choir girl standing next to the biggest and deepest organ pipe. When it played, the whole chapel had vibrated. Could the elephants, she wondered, be causing the throbbing, just as the church organ pipe had done?

Six months later she had the opportunity to find out. She was joined by the other members of the team and they returned to the Portland Zoo and recorded the sounds the elephants made. During the recording sessions they also noticed that the throbbing feeling Katy Payne had experienced on her first visit coincided with a fluttering movement of an elephant's forehead at the point where the nasal passages enter the skull. Most of the sounds they recorded, or so they thought at the time, were audible to the human ear, but unknowingly they had also recorded something else.

Back at the laboratory they analysed the audible sounds and discovered that in amongst them were other sounds that they could not actually hear. On the visual print-outs that showed when sounds started and stopped there were three times as many calls as they thought they had recorded.

Interestingly three years previously, Rickye and Henry Heffner, of the University of Kansas, had been testing an elephant's hearing range. They had been interested in the upper end of the range, which they wanted to compare with that of humans; but the results had also shown that elephants could hear down to 17Hz. They had unwittingly revealed something of great significance: that elephants can hear very low frequency sounds. And if an animal has evolved the capability to perceive sounds that low, it is almost certain that it will put it to use.

In 1983 Judith Kay Berg, of the San Diego Wild Animal

Park, had discovered that elephants have a large repertoire of sounds, including barks, snorts, trumpets, roars, growls and, significantly, 'rumbles'. Now the Cornell team had captured those rumbles on tape. The sounds recorded from the Portland elephants were at frequencies between 14 and 24Hz, and they were produced in five- to 10-second bursts for as long as 10 minutes at a time. To hear them the researchers had to speed up the tapes slightly; this had the effect of raising the pitch and brought the sounds into the hearing range of people. It was the first time that anyone had heard the secret voice of the elephant.

The sounds resembled the tummy rumbles of a hungry person, and there was speculation at one time that elephants used these sounds (produced in this way) to speak to each other. It is more likely, however, that the infrasounds are produced by the vocal chords in the usual way (a fact revealed in 1962 by I O Buss) and that, as they leave the head, they cause the skin to ripple. By their nature, it was thought, they could travel for 20km (12 miles) or more, passing through trees, bushes and tall grass. This is much further than sounds at the other end of the sound spectrum – ultrasounds – can travel. This property of infrasounds to travel great distances would better suit a creature that lives in loose herds spread out across a wide area. To determine whether the infrasounds were, indeed, an important component of elephant communication, the team went to the wild and to Kenya's Amboseli National Park.

At Amboseli, Joyce Poole and Cynthia Moss have for many years been studying the behaviour of African elephants, and have made extensive observations of their methods of communication. They have, for example, discovered that when elephants meet they need do no more than nod or twist the head or move the position of the ears or tail slightly to convey their intentions. Raising the head and spreading the ears, for example, will indicate to a subordinate bull that the signalling bull is more dominant. Even such subtle movements will cause the subordinate to give way. Indeed, most interactions between bulls are what Cynthia Moss has termed 'polite and gentlemanly': they do not charge about bellowing madly, as is often depicted in the movies.

The higher-frequency, audible components of the 'tummy rumbles' were well known to field researchers in Africa. In Cynthia Moss's book *Portraits in the Wild* she mentions that a herd will gather round the matriarch (the herd leader) in response to danger, when the audible frequencies of the rumbling sounds can be heard passing between individuals. Even by the 1960s and 1970s, field researchers were sure that these sounds were a form of communication. But until the Cornell team came to Amboseli and joined up with Poole and Moss, no analysis of these sounds in the wild had been done.

During 1985 and 1986 the researchers recorded over 1000 calls and noted the elephants' behaviour at the time the call was made. Most of the calls could not be heard because they were infrasonic, so only later could the taped sounds be speeded-up and the observed behaviour matched to them. One of the mysteries they believe they may have solved during this first series of observations is the way in which males and females find each other for mating.

Elephant herds are made up of females and youngsters; the adult males live separately. When ready to mate, bulls are said to be in *musth*: they have high levels of testosterone (the male sex hormone) circulating in the blood and can be very aggressive. They wander across the savannah, criss-crossing the home ranges of several herds of females in the hope that one of them is in heat. The chances of finding a receptive female are small, for a pregnant cow has a gestation period of two years, and a mother nurses her youngster for two years. A female is in fact receptive only for a few days every five years, and she would obviously not breed again successfully if things were left to chance. It appears she tips the odds in her favour by using infrasound.

When a female comes into oestrus, bulls very quickly

arrive on the scene – even those from quite some distance away. The research recordings of such an event showed that the receptive female emitted a distinctive pattern of low-frequency calls. Katy Payne described it as follows:

> Slow, deep rumbles, rising gently, become stronger and higher in pitch, then sink down again to silence at the end.

The sounds are made for about 30 minutes, and they are always given in the same sequence. In other contexts, such as bird and humpback whale communication, sequences of repeated patterns of sounds are usually considered to be 'songs'; and so, it seems, female elephants 'sing' for a mate.

Bulls in musth also grumble away excitedly – the so-called 'musth rumble'. On reaching a female herd, a bull's rumbling decreases and the females rumble back at him. The receptive female sings her loud song, attracts the bull, and they mate. Afterwards she gives the post-copulatory rumble. This causes the rest of the herd to join her in a very agitated state: they rumble in chorus, trumpet, and defecate in a 'mating pandemonium', as the researchers have termed it.

Not all rumbling is related to sexual activity. The research team have identified several everyday calls. When two members of the herd meet they both give the 'greeting rumble'. If one elephant is looking for another but cannot see her, she gives a 'contact rumble', lifts her ears, and turns her head left and right listening for a reply. The other elephant gives a 'contact answer' in return. And, if the herd is feeding, drinking or mud-wallowing, the matriarch will call them to order and give the 'let's-go rumble' before moving off. A hungry baby gives a 'suckle protest' to alert its mother to its need of a drink, and the mother responds with a soft rumble and allows the baby to take milk. If a youngster is in trouble and gives a distress call – often a roar or scream – the rest of the herd will come running. They rumble continuously and loudly (in fact, the loudest rumble call the researchers recorded) in order to muster more help from far and wide.

Although the researchers could match behaviour with infrasonic calls, they still had to prove that the calls initiated the behaviour and that it was not some other stimulus or channel of communication involved. This they succeeded in doing in the Etosha Game Reserve in northern Namibia in 1986 and 1987. This dry, flat region was ideal for monitoring elephant behaviour. Perched on a 6m (20ft) high tower the team could survey an area of about 130 hectares (½ sq mile). At the perimeter of their arena they placed four microphones, so that the calls picked up from any elephants entering the area could be pinpointed to a particular group or even an individual. Video recordings could be matched to the sound tapes so that sound and behaviour could be synchronized – especially valuable when the sound tapes were later speeded up to listen to the infrasound.

It was thought that in this arid environment, where the elephants continually search for water, it would benefit scattered herds to know where the best places are located. The team therefore had a waterhole in the centre of their area.

Often, several groups of elephants converged on the area simultaneously, even though there had been no visits for some time. When in the area they showed this remarkable ability of apparently silent co-ordination. Sometimes they would all panic at the same time, or they would 'freeze', as if listening for distant calls. Sudden simultaneous calling by a group of females would be followed some time later by the arrival of another group. Those already at the waterhole would greet them with a considerable degree of excitement.

Bulls, the team found, are quieter, preferring to listen to female vocal activity and for news of distant, receptive females. Indeed, in one playback experiment they discovered just how alert bulls are.

Two researchers had taken a van and had driven some distance from the study area. Those in the tower were watching two large bulls that were drinking and bathing at the waterhole and were unaware when the recordings of rumbles were to be played. Suddenly, they saw both bulls stop, lift their heads, spread their ears and move their heads from side to side as if scanning the horizon. One of the elephants swung round completely to face the direction in which the van with the playback loudspeaker was known to be, although it was far out of sight. Then both of them set out in that direction, ignoring the precious water, and stopping occasionally as if to listen. The playback tape was only 40 seconds long, and so they had no further clues; nevertheless, the two bulls kept heading in the right direction. They went straight past the van, to the relief of the two occupants, and continued in the same direction. Later, the playback crew revealed that the rumble call they had played was the 'oestrus call' of one of the Amboseli elephants. No wonder the two bulls were in a hurry!

In all the many playback experiments, the response was instantaneous. The elephants froze and listened, or returned calls. Katy Payne and her colleagues began to feel that they were beginning to obtain the evidence which would prove that elephants co-ordinate their behaviour over long distances with the aid of infrasound.

And there was further evidence. In a 10-year study of elephants in Zimbabwe, Rowan Martin had been tracking elephants fitted with radio collars. At one time it was thought that elephants wander about the home ranges in a haphazard fashion, but Martin began to accumulate evidence that showed this was not the case. Groups several kilometres apart moved and changed direction at the same time. Three or four groups, all travelling in the same direction, would follow a parallel course. Several thirsty groups would arrive at a waterhole within minutes of each other. Something was co-ordinating their activity: was it infrasound? The Cornell team and Rowan Martin joined forces to find the answer, fitting microphones to those elephants equipped with radio collars so that calls could be recorded at the same time as direction of travel is monitored.

Exciting though it is, the work does reveal a serious

problem. Katy Payne came up against it when she visited the sparsely populated Skeleton Coast Park to the west of Etosha. Here drought and ivory poachers had taken their toll of the elephant population, and by 1982 there were few left. How, then, do the few survivors co-ordinate their lives, find water and food, and find mates and companions if the remaining groups are too few and too far apart to hear each other? The silencing of elephant conversations could have far reaching consequences. It could even lead to extinction.

Protowords

One aspect of animal behaviour that has allowed scientists to gain some insights into animal communication is a creature's response to danger. In this highly motivated, life-endangered state, the alarm signs or signals an animal might give to warn its family or others in its group are likely to be distinct, but at the same time non-locatable, to the predator, and not open to misinterpretation. And these alarm calls are not accessible only to the well-equipped scientist: any enthusiast out bird watching can listen for the alarm calls of birds. The soft 'pink, pink' call of the chaffinch, heard when hawks fly overhead, is a warning call that the bird of prey finds difficult to locate. The 'rattle' of a thrush, on the other hand, is a loud rasping sound given when predators like owls are about; by this means the thrush summons other birds from far and wide to help it mob and drive away the threat. In general, the soft-edged, whispy alarms are warnings to hide, while the hard-edged, rasping sounds are rallying calls aimed at intimidating the predator.

One bird, the stonechat, has both calls – a discovery made by Peter Greig-Smith, of the Worplesdon Laboratory of Britain's Ministry of Agriculture, Fisheries and Food. He noted that throughout the breeding season the stonechat has a characteristic alarm call with two different components. One note is a thin, short whistled sound (the 'whit' note), and the second is a harsher 'chat' sound which resembles two stones being struck together – hence the name.

The two notes are used in different ways, depending on the nature of the threat. Stonechats face two types of predators: birds of prey may attack from the air, taking not only the adults themselves but also the chicks in the nest; terrestrial predators are a danger only to the nestlings. A sparrowhawk approaching the nest provokes a 'whit' call from the parent, whereas dogs and people are greeted with both 'whit' and 'chat' calls mixed together and delivered in a seemingly frantic string of loud notes. The 'whit' call has no hard edges; it fades in and out and is difficult to locate. It is given as a warning to chicks to stop begging and to be completely still, lest they attract the aerial predator's attention. If a ground predator comes close to the nest, the parent birds give the mixture of 'whit' and 'chat' calls while performing wing-flicking displays that expose white flashes from their conspicuous wing coverts, and flying erratically away from the nest. The 'whit' alerts the chicks to remain still and the 'chat' forms part of a distraction display aimed at luring the predator away from the nest.

Squirrels, like many birds, also need to know from which direction an attack is coming. Paul Sherman, of Cornell University, showed that Belding's ground squirrels – diurnal group-living rodents that live in the Sierra Nevada mountains of California – look out for their relatives. They have specific alarm calls for aerial and terrestrial predators. To those in the sky the squirrels give continuous, single note, high-pitched whistles. To ground predators the call is lower and segmented. Scott Robinson, of the University of Wisconsin, watched the way in which they used these calls and their responses. He found that they could distinguish between harmless and potentially dangerous animals approaching either from the air or on the ground. Swainson's hawks, prairie falcons, northern harriers, black-billed magpies and ospreys elicited alarm calls and escape manoeuvres. White pelicans and turkey vultures, on the other hand, might be greeted with chirrups, but the squirrels were not put to flight. Dangers from the ground come from large mammals such as dogs, coyotes, mink and badgers, from small mammals such as long-tailed weasels, and from snakes such as the gopher snake and rattlesnake. The greatest response is to large mammalian predators. When these approach, the squirrels stand on their hind legs (this is known as 'posting' behaviour), call loudly, and escape rapidly, usually into their burrows.

Robinson found that the squirrels can discriminate between situations in which a potentially dangerous predator is homing in on the colony and those in which it is just passing by. Harriers flying directly overhead are recognized as more of danger than those migrating south on the other side of a nearby river. Egrets in the air are less likely to attack than those approaching on the ground.

Escape behaviour is also related to the nature of the threat. If dogs or coyotes approach, the alarm call seems to say 'Dive down any burrow with a back-door!' The ground squirrel alarm calls are specific to a range of different predators and include what might be called the equivalent of words in human language. Indeed, hidden in those calls are descriptive 'words', which identify the type of predator, and action 'words' that tell what method of escape to use.

But of all the animals being studied by communication researchers, it is among the primates that we might expect to find signs of language like our own. Many species of Old World and New World monkeys, for example, chatter away almost continuously, show signs of a sophisticated body language, with flashing eyelids, colourful behinds and a range of facial expressions; and they have particular pieces of behaviour designed to intimidate subordinates and appease dominant bullies. Most species are social animals, with a need to communicate with others within the group or with those outside in neighbouring or rival groups. There are good reasons for their visual and verbal conversations – and good reasons, too, for scientists to study their utterances, gesticulations and postures.

Two such scientists are Dorothy Cheney and Robert Seyfarth, of the University of Pennsylvania, with ongoing work started in the mid-70s. Their interest was triggered by the developments in ape sign language and other forms of communication in higher primates. They decided that vervets – medium-sized monkeys with a greenish hue that live in troops, mainly in the open savannah areas of Africa south of the Sahara – might be useful living subjects with which to investigate whether species other than man had evolved the ability to make systematic use of signals to indicate objects around them. The attraction of vervet monkeys was that they have distinct alarm calls for different predators.

Tom Strusacker of the New York Zoological Society had made the first observations in Amboseli National Park in Kenya. He noticed that if the monkeys are foraging on the ground in an area of open bush and one sees a leopard approaching, it gives a loud bark and the rest of the troop flees to the trees. If a monkey spots a martial eagle or a crowned hawk eagle, both capable of plucking a monkey from a tree, a loud chuckle is given and all the troop look up, scan the sky, and then run for cover in the nearest bushes.

The monkeys' escape behaviour in response to the two different alarm calls is related to the hunting strategy of the two predators. Leopards hunt during the daytime, hiding in bushes, waiting to ambush monkeys passing by. A leopard can catch a monkey on the ground or in a bush but is not nimble enough to out-climb one in a tree. So the monkeys head for the trees. The eagle can take vervets both on the ground or in trees and so the safest place to avoid an eagle attack is in a thick bush.

When the troop chances upon a snake, particularly a large python, a high pitched chattering alarm stimulates the monkeys to get up on their hind legs and look carefully around them; sometimes they will mob a snake. They also have separate and distinct alarm calls for baboons, which sometimes attack young vervets, and of course for man, sometimes the most feared of all predators.

Cheney and Seyfarth decided to investigate the specificity of meaning of the alarm calls and carried out a series of experiments in the field. They located a troop of foraging vervets and planted a loudspeaker in a bush nearby. They then played back an assortment of alarm calls that had been recorded earlier in natural encounters of vervets with leopards, eagles and snakes. When the alarm calls were played the monkeys, on cue, behaved in the appropriate way and their actions were recorded on film. Back in the laboratory the films were analysed and the monkeys' behaviour noted before, during and after the playback. With an eagle alarm the troop raced for the bushes. With a leopard alarm they made for the trees. And with the snake alarm they all stood on their hind legs and frantically looked for the approaching serpent.

What Cheney and Seyfarth established is that the alarm call in itself, even without the actual predator in sight, means something sufficiently specific to the monkeys to function as if it were a word, thus dispelling the traditional notion that an animal alarm signal can be only a general manifestation of fear. It may be that an eagle alarm, for example, may refer to a recognized bird of prey, a collective noun for flying predators, or a particular escape instruction, but the important thing, as far as Cheney and Seyfarth were concerned, was that the sound represents a particular object or course of action in the external world.

Another question they tackled is whether the alarm calls are innate or learned. How do alarms develop and how is their use modified as a youngster grows up? Here there are two facets to consider: the production of the calls and the use to which they are put. The research team concentrated on the latter and made interesting observations about the circumstances in which these calls are given by juveniles.

When adults give a leopard alarm the call is almost exclusively for leopards, although lions, hyenas, cheetahs and jackals are sometimes greeted with the same call. Martial and crowned hawk eagles are the main triggers for the eagle alarm, but black-chested snake eagles and tawny eagles, both potential predators, elicit the eagle call. The snake alarm, primarily for pythons, is given also for cobras, black mambas, green mambas and puff adders. Infants, however, give alarm calls to all sorts of visitors that do not prey on monkeys – even little monkeys. They give leopard alarms to warthogs, eagle alarms to pigeons, Egyptian geese, African goshawks, grey herons, ground hornbills, lilac breasted rollers, marabou storks, secretary birds and spoonbills, and snake alarms to harmless snakes, tortoises and other things that scrabble about in the undergrowth – even mice. One small and misguided individual was seen to scream an eagle alarm to a falling leaf.

The alarm calls, even from a youngster's earliest days, are not entirely random. Eagle alarms are given only to flying objects. Leopard alarms are given only in response to animals approaching on the ground, such as wildebeest, zebra and hog. And snake alarms can be given for anything that resembles a snake, including vines, as well as for real snakes, whether poisonous, non-poisonous or constricting. Infant vervets clearly start with some kind of predisposition to give alarm calls to broad classes of predators and subsequently narrow the field down to those that matter.

Adults appear to teach the youngsters which objects warrant which alarms. If an infant gives an eagle alarm to a pigeon, the adults nearby will look up into the sky, see the harmless bird, and do nothing. But if the infant gives the eagle alarm to a martial eagle, the adults will look into the air, see the eagle, and give the alarm call themselves. The adults' response, in other words, reinforces the infant's behaviour: the infant is told either 'Ignore pigeons, they're safe' or 'You're right – these eagles are killers'.

Cheney and Seyfarth tried playback experiments with the youngsters. They waited until one became separated from its mother and then played an alarm call. The very young vervet was startled and confused, looking this way and that, searching for its mother to know what to do. Finally it ran to the safety of the parent. The adult, meanwhile, had been giving the infant many cues about the way in which it should behave. When the call was played a few days later, a frame by frame analysis of the film of its behaviour revealed that it heard the call, looked towards its mother, saw her looking upwards, and then looked up into the sky itself. A few weeks later, the youngster's response to the eagle alarm call was to look upwards automatically.

As they grow older, then, infant vervets appear to sharpen their association between specific classes of predators, the specific alarm calls, and the responses to those calls. Furthermore, they seem to have some ability to change both their alarm calls and their responses depending on the hunting behaviour of their predators. In the Cameroon forests of West Africa, for example, where vervets are hunted by humans with dogs, the monkeys' alarm calls are soft and pitched within a frequency band that matches the ambient background noise of the forest, making them difficult to detect. The monkeys respond by fleeing silently into dense bush, where humans and dogs cannot follow. In nearby savannah habitats, where vervets are not hunted by man, the monkeys climb quickly to the tops of trees, uttering loud alarm calls.

The researchers, however, were given even more tantalizing insights into vervet behaviour when they came to study a mother's response to her infant's call. It was then that Cheney and Seyfarth realized that individuals could be identified by their voices alone.

Vervets have a well-defined and complex social system. A group, for example, will have in their care a number of genetically related adult females and their offspring: aunts, cousins, sisters and daughters. The females form a linear dominance hierarchy: one female is dominant over another, who in turn is dominant over a third, and so on throughout the entire troop, down to the lowliest subordinate. This equivalent of a pecking order is reflected in the status of their young. Each infant acquires rank according to its mother's position in the troop. When two infants fight, one might scream. Its mother will hasten over to investigate. Her capacity to defeat or chase away her infant's rival is a direct function of her social status in the troop. If her infant's opponent is the offspring of a more dominant female, she will back down.

What is more interesting is that mothers can also recognize the identity of the infant that is screaming. Cheney and Seyfarth were able to test this with a playback experiment. They waited until three resting mothers were seated together and then played them the scream of one mother's infant, filming the response. When the film was analysed, the screamer's mother, as expected, responded more quickly and strongly than the other two females by looking towards her infant. What did surprise Cheney and Seyfarth was that the other females responded by turning instantly towards the mother. The mother vervet's reaction had shown that she could discriminate her own offspring from other baby vervets. The control females had demonstrated a lot more: they showed that vervet monkeys not only identify individuals but are capable of recognizing the individuals as members of different families.

There is also some degree of deception practised within the troop. Social climbing, for example, may be accompanied by 'cheating'. A low-ranking male, for instance, will sometimes give a false leopard alarm if a new male tries to join the group. They try to unnerve him, but, like a child who denies having been to the biscuit box when there are crumbs around his mouth, the vervet is seldom convincing. He will continue sounding the leopard alarm even after he has just come down from the tree signalling that there is no threat. But consistent liars, it seems, are ignored by the troop.

The troop does, however, listen for the alarm calls of other species, particularly superb starlings, which have terrestrial and aerial predator alarm calls. Cheney and Seyfarth tested the calls in playback experiments. They ignored the normal starling song, but the ground predator alarm caused many monkeys in the troop to head for the trees, and the aerial alarm made them look up. The researchers thought it showed that 'where the behaviour of another species was relevant to the vervets' survival, the monkeys' knowledge of another species' alarm calls was similar to their knowledge of their own calls'.

Mark Hauser, of Cornell University, has been looking at the way young vervets learn to listen to the birds. Like Cheney and Seyfarth, he used experimental playback of starling alarm calls to test the young monkeys. He first used the alarm given in response to a python, and discovered that young monkeys of between 23 and 84 days old recognized that the calls were of some importance and looked up, but were unsure of what to do. At about 120 days they had learned from the adults how to respond and, when the call was played, they stood on their hind legs looking for the snake, and then in panic shot up the nearest tree.

Other creatures also demand the vervets' attention. Cheney and Seyfarth played the calls of hippopotamus and black-winged stilt to vervets in an attempt to see if the monkeys were concerned if the hippo and stilt seemed to be in the wrong place. They played back the calls both in dry forest, a place where the two animals are unlikely to be found, and beside a wetland area. The hippo call was ignored whether from swamp or forest; but the black-winged stilt, whose call was an alarm call, caused some members of the troop to look up and around for signs of danger. But the monkeys were not concerned that the calls were out of context.

Likewise, vervets ignored a playback of lowing wildebeest but were very agitated by the lowing of domestic cattle. Cattle in East Africa are usually accompanied by their herdsmen, the Masai, and so the monkeys were especially vigilant when they could hear cattle approaching. Monkeys have no worries with cattle but they are very wary of people.

It was not only calls and calling behaviour that gave

Cheney and Seyfarth insights into vervet life. The *absence* of calling or alarm behaviour revealed even more valuable information. The researchers wanted to know, for example, if vervets could appreciate the presence of a predator by observing signs other than alarm calls from members of the troop or from other species or the sight of the predator itself.

Safari guides, for example, recognize that a leopard is about if they find a fresh carcass in a tree. Leopards usually hoist their kill into a tree to keep it safe from larger carnivores such as lions. Cheney and Seyfarth put a stuffed, but limp, carcass of a gazelle into a tree and watched the vervets' reaction. They ignored it. The same experiment was tried with a python trail, but again the vervets were not impressed. They were 'blind' to secondary visual cues. As a consequence, Cheney and Seyfarth considered vervets to be 'poor naturalists'. In a 1985 paper entitled *Social Knowledge in Vervet Monkeys* they wrote:

> They are disinclined to collect information about their environment when information is not directly relevant to their own survival. Vervets do not seem to know that hippos stay in water during the daytime, or that particular shorebirds do not occur in dry woodlands. These data are perhaps not surprising, but they do point out a potential difference between monkeys and human beings, who are naturally curious about much of their environment, and who engage in many activities that have little practical value to survival. We believe that these results can help us to understand the intelligence of non-human primates, and to specify more precisely how the minds of monkeys and apes differ from our own.

And to help them do this, Cheney and Seyfarth turned their attention to other vervet sounds. Vervets grunt to each other in a variety of circumstances. On moving out of their sleeping trees at dawn each morning, vervets begin to forage and, when one individual encounters another, a grunt is usually heard. One animal, for example, may approach another that is more dominant and give a grunt (a sound like someone clearing their throat with the mouth open). Twenty minutes later the same monkey may approach a subordinate and again a grunt is heard. At the edge of the forest the same monkey sits with its companions waiting to cross a clearing. The troop is nervous. The monkey grunts again. A little later, when the troop is foraging, the monkey looks up and grunts. Following his line of gaze reveals another troop of vervets on the horizon. To the human ear all these grunts sound exactly the same, but to the vervets' ears they each mean something quite different.

Traditionally, such grunts would be considered to be one vocalization, an outward manifestation of some emotional response to the environment. One monkey grunts, for example, another looks up, checks around, and sees the other troop: the single grunt in this view is not in itself like a word but more a general signal to alert other members of the troop. Cheney and Seyfarth thought this traditional view unsatisfactory. If monkeys have the ability to use different alarm calls for different predators, they argued, then why should they not have a repertoire of grunts?

Robert Seyfarth likens the human ear's inability to discriminate between grunts to the predicament facing social anthropologists studying, say, African click languages. There may be 10 different clicks, but Western observers can discern only one. A spectrograph (a machine that converts sound signals into visual ones), however, can reveal the differences; and spectrographic analysis has shown considerable variation in vervet grunts. Using playback experiments, Cheney and Seyfarth tried to discover some of their meanings.

Grunts were recorded from monkeys in a variety of circumstances. A vervet foraging alone was located, and a loudspeaker placed in a bush nearby played a grunt. Three or four days later a different grunt was played, and so on. The researchers worked through the vervets' grunt vocabulary, presenting one word each time and watching to see if it led to a response different from those grunts presented on other occasions. They were able to show that a grunt recorded from a subect approaching a dominant individual consistently evoked a different response – a turn of the head, a different posture or a different gaze – from a grunt recorded from a vervet approaching a subordinate. A vocalization perceived by human ears as one grunt is received by the monkey as many subtly different calls, each one associated with a specific and entirely different set of social circumstances. The researchers have now established the first few entries in the English-vervet dictionary.

But the dictionary, as the hippo and stilt calls, python tracks, and carcasses in trees had demonstrated, is limited: vervets have evolved only the skills they need to survive. Non-human primates, it seems, are primarily concerned only with events that take place within the troop or that directly affect the troop. Social behaviour, such as the acquisition of species-specific alarm calls, is shaped in light of this concern. Vervets have found a niche in which they can find food and live in comparative safety, and their chances of survival have been enhanced by an ability to use 'words' that not only warn of specific dangers but also trigger instant action to avoid being caught. In this they are similar to human infants, as Cheney and Seyfarth observed:

> They [non-human primates] show cognitive skills when dealing with each other, but exhibit such skills less readily in their interactions with objects. Among humans the predisposition is more subtle, but nevertheless may appear in the earliest years of childhood, when infants exhibit remarkable social skills while at the same time remaining ignorant of much of the world around them.

The researchers are beginning to get inside the vervet mind and are able to start to make comparisons with our own abilities, not by teaching the animal a man-made sign language but by understanding utterances the animals themselves are making. In short, what Cheney and Seyfarth have started to achieve with wild animals is what Herb Terrace dreamed could be gained from animals taught a human sign language. The difference is that the field researchers have been getting the animals to tell their own story in their own words.

4

Dolphins and Man

With a Little Help from Our Friends

Pliny the Elder was a chronicler of the heavens and of Earth. His *Naturalis historia*, published in 10 volumes in AD78 and a further 27 volumes after his death in the eruption of Vesuvius in AD79, were to influence scientific thinking in the Middle Ages. Some of Pliny's observations were amazingly accurate, some profound, some copied from Aristotle; but many were considered inventions or at the least misinterpretations of events he or other people had witnessed. In Book III, for example, he gave a detailed account of what some people considered to be a fanciful story about dolphins helping fishermen to catch their fish 'in the province of Narbonne, in the territory of Nîmes', then known as Gaul.

At certain times of the year, according to Pliny, shoals of mullet left the ponds of Latera and headed down a watercourse for the open sea. It would have been impossible to stretch a net across the entire entrance to the watercourse and catch the fish, for their bulk would have broken the net, so the local fishermen were confined to the shore. How, though, were they to catch the fish that were streaming down the middle of the channel? The answer was to call for the dolphins.

From the crowd gathered on the banks, a cry would go up: 'Simo, Simo!'

This was the local nickname for dolphins, meaning snubnose. And, sure enough, right on cue came row upon row of dolphins. They drove the mullet back up the channel and towards the fishermen, who encircled the mullet and dolphins with their nets. The fishermen then allowed the dolphins to eat their fill, before letting them slip over the fringe of the net and escape to safety. The following day, so Pliny related, the dolphins returned for a reward of more fish and 'bread dipped in wine'.

A similar event was reported by the second-century Greek poet Oppian. Along the Aegean shores of the island of Évvoia, dolphins were said to have cooperated with local fishermen at night. Each boat, apparently, teamed up with a dolphin and the two fished together by the light from braziers. The dolphins frightened the fish towards the boats, where the men could spear them with tridents. When the fishing was completed the fishermen shared their catch with the dolphins.

It was thought at one time that Pliny and some of the later recorders of history were a little 'colourful' in their writing and that some reports were to be taken a bit with a pinch of salt. After all, whoever heard of dolphins helping fishermen? But the narrow channel still exists at Pavalas-les-Flots, where the waters of the ponds of Mejean, Le Grec and Pérols empty into the sea from the salt marshes to the west of the Camargue. The village of Lattes (a few kilometres south-east of Montpellier) stands where the pond of Latera once filled. The mullet are still there, though not the dolphins. But the accuracy with which Pliny describes the geography seems to give some credence to his story. Could it be that men once had this rapport with a wild animal and fished with dolphins?

A clue came about 500 years ago when Portuguese navigators were exploring the west coast of Africa and had reached the shores of what is now Mauritania. By the Banc d'Arguin at the bay of that name they came upon shallow water with a rich diversity of marine life, including huge shoals of fish and even large herds of the now rare monk seal. Along the shores were the largest winter concentrations of wading and shore birds in the world. Some had come from 10,000 kilometres away, and they were here to feed on the shoals of migrating fish that swam close to the shore and on the young molluscs and crustaceans living in the sea grasses that cover many square kilometres of the clear shallow water. And here, too, the local fishermen – the Imgraguen, meaning 'gatherers of life' – fished with the help of dolphins, and it is likely they have been doing so since neolithic times.

Today, the monk seals are fewer in numbers, but the birds and the fish are still there. The area is protected, and 500 Imgraguen survive in seven villages at the southern end of the Banc d'Arguin National Park. In the Bay of Timiris, 130 kilometres to the south, they still fish in the traditional way: their companions are bottlenose and Atlantic humpback dolphins.

The fishermen anticipate the arrival of the fish in September. The mullet are heading south. When the first shoals appear in the shallows, the children splash the water with sticks to summon the dolphins. Then they watch and wait. Their very survival depends on the success of the hunt.

The first sign is a line of dorsal fins belonging to a group of hunting dolphins. Sometimes the school is over 100 strong, swimming about 20 to 30 metres out and parallel to the shore. They drive the mullet ahead of them. The fishermen rush into the water with their hand-woven nets, some wading neck or shoulder deep, and block the path of the fish shoals. As one shoal is surrounded, another net is laid further out from the first, and so on throughout the day. The fish jump about in all directions, surrounding the men with flashes and splashes of silver and showering them with shiny scales. In the panic and confusion, the dolphins avoid the nets and weave in and out of the men's legs, taking their share of the catch. It is an easy way to obtain a meal, for both the men and the dolphins. Whether it is intentional or accidental, the two appear to cooperate. The dolphins are considered to be benevolent and important to the community. They are never killed. And, thanks to the dolphins, the Imgraguen never go hungry.

Exploration in other parts of the world has revealed further evidence of man and dolphin working together. In parts of Burma in the nineteenth century, each coastal village is said to have had its own dolphin which could be summoned to help with the fishing. In the Upper Yangtse river a similar arrangement was said to have existed with Chinese river dolphins. At Amity Point in Queensland, Australia, aborigines used to tap the water with their spears and, using fishing methods similar to those of the Imgraguen, summoned dolphins to drive fish shoals close to the shore. And at Twofold Bay in New South Wales, pods of killer whales (the largest members of the dolphin family) were reported to have driven humpback whales inshore to waiting whalers. The men harpooned the humpbacks and rewarded the killer whales with the tongues and lips, which they took (according to David Gaskin's *Whales, Dolphins and Seals*) 'without interference'.

In the rivers of the Amazon basin lives the boutu or Amazon river dolphin. A freshwater dolphin, it is almost blind but finds its way about by the use of echolocation. The boutu produces clicks in its nasal area which are emitted through a large domed forehead in which is encased the 'melon' – a large, fatty organ that serves to focus the clicks into a sonic beam. The beam is directed at objects in the dolphin's path or at prey, and the echo that returns gives the boutu precise information about the composition of the object, whether it is living or dead, and how, where, and how fast it is moving.

The echolocation system is so well-developed it can even detect the thin strands of a native fishing net; and it is this ability which has given rise to another example of cooperative behaviour, which was reported by F Bruce

Lamb in 1954. Lamb was on the Tapajos river in Brazil when he discovered that the local fishermen could call up their own boutu. Sitting in one of their frail craft one night, Lamb noticed his companion tapping on the sides of the canoe with the paddle, and whistling what Lamb described as 'a peculiar call'. A few moments later along came the dolphin, herding fish into the shallows, where the fisherman could spear them. The dolphin stayed with them for over an hour.

More recently, in April 1988 and February 1989, Karen Pryor and Jon Linbergh from North Bend, Oregon and Scott Lindbergh and Raquel Milano from Brasilia, visited the town of Laguna, between Porto Alegre and Florianopolis in southern Brazil, and discovered wild bottlenose dolphins cooperating with saltwater fishermen. Like Pliny's experience in the south of France, the catch is mullet and the action takes place at the entrance to a series of interconnecting brackish-water lagoons. The one nearest the sea is called Lagoa Santo Antonio, and it is here that 30-40 fishermen ply their nets every day of the year, barring storms and a couple of lean months during midwinter (July and August).

Local records, according to a report in the January 1990 edition of *Marine Mammal Science*, show that this unusual method of fishing started in about 1847, and it is not just for fun or for the table at home; the catch is sold at local markets. The fishing follows a set pattern, almost a ritual adhered to both by man and dolphin.

The fishermen, each carrying a circular net edged with weights, wade out into the water and stand in a line parallel to the shore waiting for the dolphins. The water is murky so the men must rely on the dolphins to know when to throw their nets. The dolphins appear first some distance away but slowly swim towards the line of men. One by one they submerge and swim closer. Suddenly, a dolphin will emerge from the water just in front of a fisherman and turn on its side. This is the signal for the fisherman to cast his net. If the dolphin does not perform a roll the fisherman will not cast.

In fact, the dolphins seem to lead the entire hunt dictating, by their position offshore and their behaviour when herding the mullet, where the men should stand and when they should cast their nets. Their reward, if the fishing is good, is 10-12kg of mullet a day which they are able to catch with ease in the confusion among the nets.

Dolphins to the Rescue

Remarkable behaviour on the part of dolphins has not been limited to cooperative hunting. Down the centuries, many stories have been told of dolphins saving people's lives. Maori legend, for example, has the *taniwha* that saved people from the sea. It is thought that this legendary creature is based on the dolphin.

In Greek folklore, the seventh-century BC poet Arion was supposed to have been saved by a dolphin. He was thrown into the sea by pirates who had hijacked his ship, and was later carried ashore by an obliging dolphin. The event is commemorated on a third-century Greek coin which can be seen in the British Museum, London.

More recently, there have been tales which appear to have the ring of truth. In January 1989, three teenage surfers in Australia claimed they had been saved from a three-metre-long shark (possibly a great white – the largest predatory fish in the sea) by a school of dolphins.

The three were surfing among the dolphins off the isolated Halftide Beach, south of Evans Head in New South Wales, when the dolphins became agitated and began milling about below the surfboards. Then, suddenly, the shark attacked. In true great white shark fashion, it approached its victim from below and behind and slammed into the surfboard taking a chunk out of the board and the boy. Blood streamed into the water, and the boy struggled and lashed out in an attempt to discourage the shark. At this moment the dolphins began to splash the water and ram the shark, driving it away while the boys limped back to the safety of the shore.

The experts say that the incident does not support the notion that the dolphins were helping the boys. More likely, the animals were defending themselves from the threat and it was simply coincidence that the boys benefited from their action. There is no evidence of 'altruistic' behaviour, say the scientists. But how right are they?

Not far from Delagoa Bay on the Mozambique coast, a 20-year-old girl from Pretoria was trying to swim to the shore for help after her boat had capsized. She had cut her foot during the accident and the blood in the water attracted four sharks. Suddenly, two large dolphins appeared. They not only chased away the sharks but also came alongside the girl and supported her as she swam to a shipping buoy. She was exhausted and, at first, was unable to climb onto the buoy. The dolphins stayed and supported her until she was able to haul herself from the water. A look-out on a passing tanker spotted her and radioed for help.

Newspaper reports in 1978 also highlighted a dolphin-helps-man-incident off the South African coast. People fishing from a small boat were caught in a pea-souper of a fog and lost their sense of direction. Suddenly, a school of four dolphins gathered around the boat and began to push it this way and that as they steered it through a series of dangerous rocks and reefs. After an hour and a half, the dolphins nudged the boat into safe waters – then left as suddenly as they had come.

In Horace Dobb's *The Magic of Dolphins* the author tells of a conversation with the son of Zachious Benga, a hunter turned fisherman who founded the village of Mama Beach in Sierra Leone. One day Zachious was out at sea in his dugout canoe with a group of other fishermen from the village. In the evening, they began one by one to leave the fishing grounds to return to the village. Zachious was one of the last to leave. To his horror his boat began to break up – the wood had dried in the sun and cracked – and he ended up clinging to one of the large chunks left floating at the surface. When he did not return that night the other fishermen thought he had fallen foul of sharks. But Zachious was determined to survive. He lay along the top of the remains of the canoe trying not to fall asleep. and throughout the night he could hear splashing all around him. He thought it was sharks waiting for him to fall off. But, come morning, he discovered that his companions had not been sharks but dolphins, which continued to circle the wrecked canoe for a while, and then disappeared. For the rest of the day he was on his own, desperately tired and weak from the hot sun overhead. As the darkness of the second night descended, the old man was afraid that he would slip into the water, but just then the dolphins reappeared around him and splashed water at him to keep him awake.

Having survived another night, Zachious was resigned to the fact that he would not be rescued, and was about to resign himself to his fate when a trawler spotted him and took him back to their home port, a long way up the coast from Mama Beach. Meanwhile, back in the village, the fisherman's family and friends were in mourning. There was no way a message could be got to them. According to local Christian custom, the final ceremony associated with a death takes place 40 days after the event. Imagine, then, the shock and then delight experienced by the Benga family, all of whom were gathering from far and wide to mourn the old fisherman, when 30 days into his period of mourning, Zachious walked down the village street alive and well.

In another incident recalled in Hans Hass's *Conquest of the Underwater World* and Maurice Burton's *Just Like an Animal*, a solitary dolphin was reported to have saved the life of a woman in danger of drowning off the Florida coast in 1943. She was the wife of a lawyer and had been bathing alone off a private beach when, finding herself in waist-deep water in a fast current, she became paralysed with fear. She swallowed a considerable amount of water and was near to drowning when she received what she described as 'a tremendous shove'. At that point she was spotted by a man on the shore. He saw her almost lifeless body floating in the water, when a large glossy animal he at first took to be a shark nudged her to the shore; shortly afterwards he saw a dolphin jumping about in the water. More recently, a scuba diver in difficulties off the Cornish coast was apparently supported and then helped back to the diving tender by a dolphin – an event witnessed by other divers and the back-up crew in the boat.

In *The Dolphin's Gift*, Elizabeth Gawain recalls a conversation with a Californian diver, Joe McCord. He had been diving off the Pacific coast of Mexico when he developed stomach cramps. His air was rapidly running out and he was unable to unfasten his weight belt in order to float to the surface. Desperately he called out in his mind 'somebody, something, help me!' as if trying to summon up some 'vague power', as he described it.

With just ten minutes of air left, he felt a nudge behind his arm. At first he was terrified for he thought it was sharks homing in for the kill. But then he saw an eye:

> The most marvellous eye I could ever imagine. I swear it was smiling. It was the eye of a big dolphin. Looking into that eye, I knew I was safe.

And the dolphin took McCord into shallow water, almost beaching itself in the process. McCord took off his diving gear and returned to the dolphin which waited in shallow water, as if checking that the man was alright. Some way offshore, a large school of about 100 dolphins seemed to be waiting for their colleague. McCord and the dolphin played for some time, the dolphin carrying the man on its back. The encounter ended after the man and the dolphin stayed motionless and looked at each other in silence. Then the dolphin made a noise and raced off to join the rest of the school.

Rescue dolphins, though, have not always been welcomed by shipwrecked sailors. *National Geographic*'s Edward Linehan relates the tale of a bunch of US Air Force flyers during World War II who had been forced to ditch in the

South Pacific and had to drive away their friendly dolphin: it was pushing their rubber dinghy towards a Japanese-held beach!

Reach Out for a Dolphin

Dolphins have not only come to the rescue of people in trouble in the sea; they are also contributing to medicine as nature's psychologists. In an experiment in the Soviet Union in the mid-1980s, pregnant mothers were encouraged to swim and dive before they came to term. They then gave birth to their children actually in the water. The benefit, according to Igor Charkovsky, the supervisor of the experiment, was that the children developed more quickly than children born by conventional methods. The 'water-babies' began to walk at four months old and could swim several kilometres at the age of six. In a contest with French diver Jacques Maillol, a world-record holder for diving to 100 metres without breathing apparatus, water-born children played on the bottom of a pool long after he had to come up for air.

Charkovsky worked with Academician Pyotr Anokhin, a Soviet scientist who believes that the human foetus has a psychology all its own and is capable of learning while still in the womb. As part of an experiment, Charkovsky introduced pregnant mothers to dolphins. Their presence, he thought, had a soothing effect on people. He also believed that a telepathic link once existed between man and dolphin, and he wanted to prove this to be the case.

One of Charkovsky's patients swam in the sea with dolphins for the first few months of her pregnancy. When her child was six months old, she returned to swim with the dolphins, and both the mother and Charkovsky were in for a big surprise; not only did the dolphins appear to recognize the human baby, but also he, in turn, reached out for them 'as if they were friends of old'.

At the Dolphin Research Center in Grassy Key, Florida,

Tuesday is a rather special children's day: the children are mentally handicapped, and they come to swim with the dolphins. It is psychologist David Nathanson's way to encourage them to speak and remember things. At the Center, where the children can swim with and cuddle dolphins, they learn much faster than in the normal classroom environment; Nathanson believes that the dolphins improve the attention of the children. This enables them to learn more quickly than usual. Learning involves two basic steps: reception of a stimulus, and processing the information contained in the stimulus. Mentally handicaped children have more problems with the first step and it is here that the dolphins help.

In a study in the summer of 1988, Nathanson introduced six children, aged from two to 10, to the Center's dolphins – three children had Down's syndrome, another had hydrocephaly, 'water on the brain', one six-year-old had speech and memory problems as a result of contracting meningitis just after it was born, and the 10-year-old was both physically and mentally handicapped. At each session the children were shown large white boards, each with a single-syllable word or a picture. If the child identified the opening consonant or managed to say the whole word it was rewarded with feeding, kissing, hugging or swimming with a dolphin. Sometimes the dolphins help with the lessons. A picture of an easily recognized object, such as a dog or a car, is put into the pool and any one of the six dolphins taking part in the project will bring it to one of the children. If the child correctly identifies it and pronounces the word properly, he or she gets to swim with the dolphins. The children need little encouragement to succeed; the dolphins are the tops – the children prefer a hug with a dolphin to a hug from their human teacher.

Nathanson was overjoyed with the results. In *Focus on the Sea*, published by the Florida Oceanographic Society, he wrote:

> The results of the research were startling. Every child responded two to ten times more working with dolphins!

> The reward of interaction with dolphins was more motivated than I had anticipated.

According to Nathanson, the dolphins seemed to sense there was something special about the children. A six-year-old quadriplegic boy was placed on the edge of the pool with his legs dangling in the water and a dolphin called Little Bit (one of the ones that took part in the *Flipper* TV series) sought him out and nuzzled up to him. The dolphins seem to recognize that the children are more helpless than the people with whom they usually swim and treat them much more gently. Nathanson saw that this work could go some way to helping the cause of dolphin conservation:

> A picture of dolphins helping people is a powerful image. Perhaps we ought to use such images to make sure that people and governments properly protect and manage the seas. We cannot take for granted the irreplaceable resources found in nature. By helping people, dolphins may ultimately be helping themselves and the other creatures who inhabit the world's aquatic environments. After all, who would not support protection of dolphins after they've seen Little Bit gently push a crippled child to the surface of the sea?

Similarly, at Dolphin Plus (Key Largo), Betsy Smith of the Florida International University, Miami, tested her belief that dolphins could help autistic children who had not gained a great deal from more conventional treatments and teaching. She believed that children in contact with animals would become more sociable and communicative.

After eight days at Dolphin Plus she found that autistic children did, indeed, improve after being with the dolphins. Whether this was due to the dolphins or simply because of the therapeutic effect of the water, Smith is not certain, although a control group that went swimming on a nearby beach did not do so well. Moreover, when she talked later with the parents of the children who swam with dolphins it became clear that there were significant beneficial long-term behavioural changes in that group.

Dolphins have also been brought in to help cancer patients. For some time, some terminal cancer sufferers have succeeded in reducing the size of their tumours simply by imagining that the cancer is something evil and that the evil can be overcome by the forces of good – much like St George slaying the dragon. Stephen Jozsef believed that dolphins might help in that fight. He works with terminal cancer patients at Living From the Heart, an institute in Parker, Colorado. Joszef tested this belief by flying 15 cancer patients across the United States so that they might swim with the dolphins at the Dolphin Research Center. Dolphins, Joszef thinks, live in 'an alpha state'. This is a state of the brain that occurs when we close our eyes and relax but do not actually go to sleep. This 'alpha rhythm' is the goal of those who aspire to meditation and self-control. Joszef believes that by achieving the alpha state people could achieve some degree of self-healing, and that swimming with dolphins might help patients to tune into their own healing powers. By taking blood samples and monitoring brain-wave patterns before and after the contact, Joszef hopes to present the medical establishment with the necessary scientific data.

On a slightly different plane, Rose Farrington, creator of the New Games movement, was inspired to seek a new direction in her life's work by three captive dolphins at the Steinhart Aquarium in San Francisco. At first she thought it cruel that they should be deprived of their freedom, but after standing in front of their tank for some time, she began to feel as if the dolphins were actually talking back to her. 'Words', she told author Elizabeth Gawain (whose life was also changed by dolphins, as told in *The Dolphin Gift*), 'flowed back into my mind'.

From the dolphins she discovered the importance of play as a vehicle for love and co-operation, and during this experience Farrington watched the three dolphins playing with a short piece of orange yarn. During her musings, the dolphins seemed to be telling her that bringing the peoples of the world together in play was a way in which peace might be achieved. With the support of the producers of the Whole Earth Catalogue, Farrington organized the first Whole Games Tournament at Marin Headlands, north of San Francisco. In following years, she lectured and organized events around the world – and it all started from three dolphins and a piece of yarn.

What we make of such claims and events, I have no idea, but down through the centuries people all over the world, from different races, religions, creeds, cultures and backgrounds claim to have had contact and even some form of communication with these animals. There is no doubt that any contact with dolphins is special.

In the United States that 'contact' has been commercially exploited. At the Dolphin Quest of the Hyatt Regency Waikoloa Hotel in Hawaii, one of the largest captive dolphin facilities in the world, for example, punters can spend $65 (£38) for the chance to swim for about 10 minutes with dolphins. More 'encounter centres' already exist, including four in the Florida Keys area, and more are seeking permission to set up these money-spinning facilities. Whale conservationists are understandably worried about the proliferation of this kind of centre. Most of the cetacean occupants must be caught in the wild. There is also the risk of dolphins succumbing to human diseases.

There are, however, centres like Dolphin Plus and the Dolphin Research Center mentioned earlier, that put their dolphins first and people second.

At Dolphinlab, headquarters of the Dolphin Research Center, the whole thing is taken very seriously. Human-dolphin encounters are preceded by a 30-minute educational session in which visitors are introduced to the concept of the dolphin encounter, given a background talk on the work of the facility and taken on a tour. After the swim there is another workshop and discussion about the event. The Center is a non-profit corporation that puts a high priority on education and research. Indeed, man-dolphin communication experiments have been a feature here, as well as courses in dolphin behaviour, training, and husbandry, reef and sub-tropical coastal ecology, and marine ornithology. The Center has also provided holidays for stressed dolphins that have lived in tanks, and have taken in any that have stranded nearby, including spinner dolphins, a large male bottlenosed dolphin, a pod of pilot whales, and a beached baby sperm whale that was kept alive for 12 days.

As for the permanent inmates, they seem to enjoy human-dolphin encounters as much as people do. It has even been suggested that the dolphins think that the Center owners are capturing people for the dolphins' enjoyment. The notion gains plausibility from the fact that the dolphins are not in tanks but in large pens separated from the sea by low fences. If they wanted to, they could easily jump the fences and make good their escape; but they don't.

On the other side of the Atlantic, thousands of dolphins pass through the Strait of Gibraltar each year. Waiting for them with his catamaran is Mike Lawrence, and passengers who join him at Sheppard's Marina in Gibraltar lie down on the deck and touch the wild common dolphins that come to swim between the twin-hulls of the boat. Once contact is made many dolphin-watchers seem mesmerized: it appears that something extraordinary passes between man and dolphin.

That special 'something' can be hinted at by recounting two stories. The first was the occasion when Antonietta Lilly, wife of the famous dolphin researcher, met a white whale or beluga at a new cetacean research facility at San Diego. The incident is described in the prologue to John Lilly's book *Communication Between Man and Dolphin*. Captivated immediately by the belugas' gleaming white bodies, flexible necks and variety of facial expressions, Antonietta decided she would like to swim with them. She had an overwhelming feeling that she was 'being perceived by an immense presence', but summoned up enough courage to enter the tank. As the whales circled around her, interrogating her with their sonar, she made some sounds underwater that attracted the whales closer. At that point, the contact was interrupted by some researchers wanting to take some recordings. Reluctantly, Antonietta left the water and waited for the experiment to finish.

As she sat on the side of the tank, a beluga popped its head out of the water directly in front of her, pursed its lips (as only belugas can), and squirted water at her face. Antonietta responded by taking a mouthful of water and squirting it back at the whale. From that moment on the two played together and human and whale were as one. 'This whale's invitation to share her world,' wrote Antonietta, 'gave a glimpse through a cosmic crack between species.'

This same fissure was opened a few years later under quite different circumstances. Unlike the cancer patients who have undergone dolphin contact therapy and gained some degree of remission, a 15-year-old girl from North Carolina was not so fortunate. Her treatment was halted after her cancer spread. She had wanted to be a marine biologist. She loved the sea and everything about it. She also had fallen in love with dolphins after watching the TV series *Flipper*. Her greatest ambition was to swim with these beautiful creatures, and thanks to the Children's Wish Foundation of Atlanta her wish came true. She was taken to the Dolphin Research Center in Florida.

Supported by a life-ring and helped by her father, the frail teenager floated in the water and spent over threequarters of an hour with two bottlenose dolphins – a male called Nat and a female, Turci. The two animals were chosen for their gentleness, and they pulled the young girl round their pen, jumped over her and allowed themselves to be hugged and kissed. For a few brief moments she had been at one with representatives of another species. Three days later the girl died.

Riding Wild Dolphins

The dolphin's affinity for people is explained in Greek

mythology by a story about Dionysus, the god of the vine. Dionysus once came down from the heavens and, disguised as a mere mortal, travelled on a ship. The crew, thinking they could make a little money on the side, seized him with the intention of selling him as a slave. Gods, of course, have 'second sight' and Dionysus was able to foil the plot against him. He decided to teach the crew a lesson and changed their oars into snakes. The terrified sailors leaped overboard in panic. Immersion in the Mediterranean brought about a rapid change of attitude, and they were sorry for what they had done. Dionysus relented and changed the men into dolphins, so they would not drown. Ever since, dolphins have had a special relationship with man. (The story is told in pictures on the famous Dionysus cup of 540BC, now in Munich.)

Legend, however, has been mixed liberally with reality. Aristotle (384-322BC), for instance, records dolphins having 'passionate attachment to boys in and about Taras, Karia and other places'. Greek and Roman coins bear images of men and boys riding or being rescued by dolphins.

Pliny the Elder has a splendid story which he gleaned from the chronicles of Mecaenus Fabianus and Fluvius Alfius. He tells of a dolphin in the Gulf of Lucrinus, not far from present-day Naples, which was summoned each day by a schoolboy, who befriended the dolphin by offering him bread and other foods. The dolphin had probably been trapped when the bay had been sealed off from the sea by a kilometre-long dyke. The dolphin, deprived of companionship with others of its kind, welcomed the attentions of the boy. Once the relationship had been established, the boy would call for the animal at noon to take him from Baianum to Puteoli. The boy, it is said, rode to school on the back of the dolphin!

Some of these ancient accounts were no doubt exaggerated in the retelling, but the fact that dolphins have an affinity for people and their ships cannot be disputed. Pliny the Younger describes a dolphin that came to the beach and frolicked with bathers at the ancient Carthaginian town of Hippo (now known as Bône, on the Algerian coast). In a letter to his friend Canius Rufus, Pliny describes how Hippo was on the side of a lagoon which was connected to the sea by a narrow channel. Strong currents formed in the channel and young boys would see how far out to sea they could be carried. One boy had gone out quite a way when he found a dolphin underneath him which proceeded to carry him further out to sea on its back. The boy was terrified, but after a while was returned to his friends by the dolphin. During the next few days the dolphin came back and the boys became friendly with it, touching and playing with it. It even allowed the original passenger to ride with it once more. Eventually the animal became known for miles around, the local populace took umbrage at the influx of tourists, and dolphin riding was banned henceforth.

In the late nineteenth century, a dolphin in New Zealand waters enjoyed the unusual distinction of an official protection order. He was Pelorus Jack, a Risso's dolphin (although some believe it to have been a large bottlenose dolphin) that escorted ships in the vicinity of the Cook Strait, which separates North and South Islands. From 1888 until 1912, according to an account in Ronald Lockley's book *Whales, Dolphins and Porpoises,* the dolphin joined every ship that entered the south side of Pelorus Sound (east of Tasman Bay on South Island) and appeared to guide them for a distance of about 10 kilometres until they passed through the tide-race at French Pass. Pelorus Jack regularly rode on the bow-wave of the ferry that plied the route between Wellington and Nelson and became so popular that a unique order-in-council was issued by the Governor General to protect the animal. Indeed, so engaging was Pelorus Jack's behaviour that the president of the Linnaean Society, Sir Sydney Harmer – an authority on whales and dolphins – wrote: 'we may have to review our incredulity in regard to the classical narratives of the friendliness of dolphins towards mankind.'

Further support for this suggestion came again in New Zealand in 1955. The place this time was Hokianga Harbour, Northland and the principal performer was Opo, short for Opononi Jack because she (not he, as was first thought, whence the name 'Jack') was first discovered near a sandy beach at nearby Opononi, nudging small boats. She became so used to people that she would allow some to touch and stroke her. It was thought she was the orphan of a dolphin which had been shot by a local boy and was therefore seeking company. In fact, Opo struck up a particular 'friendship' with 13-year-old Jill Baker, who was able to ride on the dolphin's back. In Anthony Alpers's *Dolphins: the Myth and the Mammal,* Jill Baker describes how the dolphin became her special friend:

> I think why the dolphin became so friendly with me was because I was always gentle with her, and never rushed at her as so many bathers did. No matter how many went in the water playing with her, as soon as I went for a swim she would leave all the others and go off, side-by-side with me.... On several occasions when I was standing in the water with my legs apart she would go between them and pick me up and carry me out a short distance before dropping me again.

Jill's experience certainly gave credence to those ancient tales of people riding on the backs of dolphins. But this story had a sad ending. The local authorities became concerned that Opo would be injured or killed by vandals and issued another special order-in-council. On the day it was to come into force, the dolphin died. She had become trapped in a crevice and left behind by the receding tide. Opo had been so popular that she was buried next to Opononi village hall and a special statue was erected in her honour.

A happier fate awaited another trapped dolphin, who was left behind on a sand bank by the retreating tide on the Isle of Man. Donald, as this dolphin was called, was saved by his human friends, who formed a bucket-chain and poured water over him until the sea returned. Donald was a bottlenose dolphin that spent much of his time with bathers and divers off the coast of Man between 1972 and 1974. Why Donald should have welcomed the attentions of people is hard to understand. He had been shot by some mindless youth and later had collided with an outboard motor. The scars, however, made him easy to identify.

Donald was brought to my attention by an Isle of Man diver, Maura Mitchell. I first spoke to her after her encounter with a school of over 70 basking sharks (the second largest fish in the sea) off the Calf of Man in the summer of 1989. She told me of an even more exciting experience she had had many years previously.

Maura was living on the mainland at the time, but had returned to her home island to dive. One of her underwater companions was Mike Bates, of the Marine Biological Station at Port Erin, and he introduced her to the world of the friendly dolphin. Maura first met Donald when diving one day at Derbyhaven (east of Castletown). The divers abandoned the planned dive, discarded their aqualungs and, taking snorkels and face-masks, they swam with Donald. Much to the disappointment of the male divers, Donald took far more interest in Maura, who, it seemed, had an advantage over Donald's other visitors. Having been brought up on a farm with animals, particularly horses and ponies, Maura felt she had an empathy with the dolphin.

> I spent most of my childhood on the farm just messing around with animals. I relate to animals very well. If people have a horse that has a cut or something and they can't catch it to dress its leg, they yell for me; and I can catch the uncatchable. It didn't matter whether it was dogs or cats, pigs or chickens or anything, I just seemed to get on with them. When I met Donald I realized instantly I could start on a different level. With most animals you have to put yourself in their mind – it's very difficult to describe, but you have to feel for them and not give them human emotions and so on. So, on my first meeting with Donald I recognized immediately that I didn't have to go through the process of

> being accepted by the animal. He was instantly interested in whatever I did, whether it was changing shackles on moorings or crab hunting. He was fast to cotton on. You could show him, like an alien, what you were doing and he seemed to understand. On one of the first dives with him I showed him how human hands work, and although this may seem far-fetched, I showed him how each finger wiggled and how I could pick up stones and drop them, how I could manipulate kelp, hold my camera, and push things – and he had his beak against my hands while I was doing them, and nodded very rapidly at everything I did. But, once he'd learned that, you couldn't do it again because he wanted something different. You had to show him different things all the while. It wasn't me studying him, it was him studying me.

And, all the time Donald and Maura worked and played together, the dolphin would scan with his sonar, making the 'creaky-gate' noise that Maura could hear under the water.

Donald had several games he liked to play. Indeed, people knew when Donald was about because small boats moored in the harbour took on a life of their own. He took great delight in pushing a boat to the end of its mooring and letting it go. Another favourite pastime was spinning vacant mooring buoys, an activity that was nearly his undoing when he was caught by the tail between twin ropes from one of the buoys. He also liked to tow a small boat by pulling on a deck quoit attached to a rope held by someone in the boat. There were also underwater games.

> He watched me pushing a ball underwater and it would pop up. He stood on his head and used his tail to push it under.

On one occasion, Maura tried an experiment.

> I got a sheet of perspex and put a spider crab, first on top and then underneath. Donald hung above it and scanned it, and quickly realized, when I showed him, how to get the crab when it was underneath. On another occasion, one of the

> divers was 'flatty-bashing' (spear fishing is banned around the Isle of Man but hand-spearing of plaice is permitted) and Donald copied him by pinning a flatfish with his beak, as if to say 'I can do that as well'. He was also very good at finding things, like lobsters, for you. He just hung above it and quivered.
>
> He had very good body language, which was very expressive, like an excited child. He almost squirmed with delight, with his head down and his tail up in a hump, and he quivered, almost as if he was laughing. He had a devilish sense of humour.

But Donald was not all sweetness and light. When divers were busy working and ignored his requests to play, he would pin them to the bottom of the sea and wouldn't let them go. Off the Isle of Man three professional divers and a novice were treated in this way. It was, to all accounts, most unnerving, and the newcomer to diving was so scared by the persistence of the 3.6m (11ft 10in) dolphin that he never dived again. And one diver exploring a wreck had Donald grab him by the ankle. He didn't know Donald and his liking for games, and he came to the surface mighty fast.

Sometimes Donald got angry. He showed his displeasure by tail-slapping – leaping up and down and banging his tail on the water surface. He became angry if people disturbed him when he wanted to rest. He liked to lie alongside boats, particularly the old-style lifeboats or bilge-keel yachts. When dolphins sleep they 'shut down' one side of the brain at a time (sleep occurs every 20-40 seconds between breaths and amounts to about 120-140 minutes a day for each side of the brain), and Donald seemed to find security in placing his sleeping side next to a boat. After a while, he would face the other way and let the other half sleep. And woe betide anyone who interrupted his afternoon rest.

Maura was never at the receiving end of his wrath, but she did upset him once. A newspaper photographer was taking pictures of Maura with Donald but ran out of film. Maura suggested he use her camera which was down in the cabin of the boat.

> I tried to describe where it was, and I was a little bit overwound because he didn't really understand what I was saying. I was waving my hands around and pointing and Donald mistook this. He thought something was wrong. He put his beak in my tummy and propelled me backwards about ten yards, until I patted him on the head and reassured him that there was nothing wrong in the boat that I needed to be protected from, and then he was alright again.

Maura felt that her conversations with Donald were two way – he talking in body language, she in sign language.

> He understood what I was on about and I understood what he was saying. I just accepted him as another intelligent being, just as you'd show somebody from another culture who couldn't speak English – you'd just work in sign language.

There was also, she thought, an element of what might be called extra-sensory perception. Acknowledging that one should try to be scientifically objective when describing such things, she nevertheless insisted that something unusual happened when the two of them were together:

> It just got spooky. Things happened too often to be coincidence. When you were going to do something, he'd anticipate. It happened a lot when we working with him filming. He'd do things on cue without you explaining. There was one occasion which I didn't witness but was told to me, and which the group of divers involved couldn't explain. They had been playing with him and were about to go on a dive. Sitting in the inflatable, they said 'Let's go on our dive now', and one said 'Oh no, I can't, I've lost my depth gauge – I've dropped it somewhere here.' Whereupon, Donald disappeared below and reappeared a few minutes later with the gauge in his mouth. Now, you explain that!

And that story, according to Maura, was from a boatload of perfectly sane divers. Donald was also polite. He was frequently offered fish by passing fishing boats, but he didn't really like them: he preferred to catch his own. He accepted the gifts, but did not eat them:

> He started stacking the fish on top of a 'can' mooring buoy. So, when the fishermen came along on the next occasion, they put the fish on the can buoy, thinking that was what he wanted. Instead, he came along and just swiped the whole lot off with his tail. He was saying 'Stupid things, can't they see I'm trying to give it to them back.'

Early in 1975, dynamiting to improve the harbour at Port St Mary drove Donald away. But he soon reappeared at Martin's Haven and Skomer Island, to the south of the Isle of Man, where he gained the name 'Bubbles'. After a spell at Milford Haven in Wales, as 'Di the Dolphin', he turned up yet again – this time he was called 'Beaky' – at St Ives and Mousehole on the Cornish coast.

Donald became famous throughout the British Isles and in many other parts of the world, much as a result of the interest of Horace Dobbs and the publication of his book *Follow a Wild Dolphin*. Maura Mitchell introduced Dobbs to Donald and the encounter changed the course of his life. He had been visiting the island on business, and his host was to be away for the weekend. Knowing he was a diver, Maura was asked if she would like to entertain him.

> I said to him, 'I've got a rather unusual friend, would you like to meet him?' He was absolutely captivated and he came back again and again.

In fact, Dobbs was so captivated he abandoned a career in medical and veterinary science and devoted his life to the study of dolphins, particularly those that seem to seek the attentions of people. Dobbs set up the International Dolphin Watch, but Donald was just the beginning. There was

Dobbie in the Red Sea (who was murdered), and, in *Tale of Two Dolphins*, Dobbs tells of Percy of Portreath (Cornwall) and Jean-Louis of Brittany. But there was one event that seemed to Dobbs to encapsulate the relationship between man and dolphin.

One summer's day in 1984, Tricia Kirkman was with Dobbs and Percy. She was a non-swimmer but was determined to summon up courage to enter the water with the dolphin. Dressed in a wet suit, with mask and snorkel, she slipped quietly into the water. After ensuring that she was safe, Dobbs climbed back into the inflatable, changed the film in his camera and was just about to re-enter the water when he saw something that was to become permanently etched into his memory:

> Tricia was floating on the surface with her arms straight down and her hands resting very lightly on Percy's head. The dolphin was swimming very slowly forwards so that the two of them were gliding through the water.... It was a picture of great beauty, peace and harmony, one of those incidents in life which last only briefly, yet are extremely significant. Here was a woman who could not swim, floating across the deep sea, propelled by a wild dolphin.

Since Donald, Percy and Jean Louis, many other incidents with friendly wild dolphins have been reported. There was Seamo in Cardigan Bay, Fungie off the Irish coast, Sandy of San Salvador, and many, many more.

'It's as if Donald had told all the other dolphins,' thought Maura Mitchell, 'that people aren't all bad.'

And that sense of caring about dolphins was illustrated recently in north-east England. Freddie, a dolphin 'resident' in Amble harbour, Northumberland, had a narrow escape in December 1989 when Royal Navy divers wanted to blow up an old World War II mine. Four amateur divers lured Freddie away from the area and kept his attention while the military exploded the 500lb of high explosive. Freddie, who had been at Amble for three years, was safe.

Are these dolphins – known as 'friendlies' – actually seeking human friendship or is it a by-product of some normal dolphin activity? According to Dennis McBrearty, of Cambridge University, who has been co-ordinating dolphin sightings around the British coast in the Dolphin Survey Project of the International Dolphin Watch, dolphins in British waters occur in groups numbering from two to 25, and there are many solitary dolphins. It is inevitable, he thinks, that some dolphins are going to take up territories in bays and coves not far from human habitation. The chances are that humans will take the initiative and make the first contact.

This was what happened with Fungie, who has now taken up residence in Dingle Bay on the west coast of Ireland. For years Fungie had stayed clear of divers, but eventually two local divers, John O'Connor and Ronnie Fitzgibbon, gained his trust and were allowed to swim with him and even touch him. And over the past couple of years the dolphin has become a national TV celebrity. Local fishermen are pleased that Fungie is there. On this exposed coast, open to the power of the Atlantic Ocean, fishing can be dangerous. As long as there is a dolphin about, according to local folklore, nobody will drown. Fungie is also good news for local commercial interests: he draws crowds of tourists.

And in the little Spanish town of La Corogna, Nina similarly entertained thousands of tourists. She had favourites among the many divers who came to swim with her, and accepted fish only from them. A local by-law was introduced preventing outboard motors from being used near the dolphin and prohibiting the use of nets in the bays which she frequented. Unfortunately, Nina was not so lucky as Freddie and Fungie: she was the victim of an underwater explosion.

Then there was Dolly of Florida Keys. She swam up a creek each day to accept fish and play with a family who had a waterside property and dock some way from the sea. Dolly was taught to recover coins. A handful of brown and silver coins would be thrown into the water and she was asked to recover just the silver ones. The water was very murky and humans trying to do the same trick failed completely. Dolly, on the other hand, returned unerringly with the correct 'change'. (It turned out that Dolly was not totally wild, but was a reject from a US Navy training programme and had been missing her human companions. Eventually, she was taken back into captivity and lived out her days in a dolphinarium.)

During the making of the movie *Flipper*, the dolphin researcher John Lilly was able to confirm the boy-rides-dolphin stories. He was asked by film producer Ivan Tors if the young hero of the film could ride a wild dolphin. Lilly thought it could be done, and so the theory was tested in a strip of water on a Bahamas beach. Two female dolphins and a youngster were placed in a large enclosure next to the beach and Ivan Tors's two sons accompanied Lilly into the shallows. On the first day, the three intrepid humans encouraged the dolphins to come closer and closer until the three animals allowed themselves to be stroked and petted. On the second day, they needed little encouragement and were quick to greet the children. The mother dolphin even allowed the boys to climb onto to her back, grasp her dorsal fin, and she would take them for rides out to deeper water. She would dive below the surface, but only long enough for the boys to hold their breath, and returned to the surface; then, she brought them back to the shallows.

So could those Greek and Roman legends be true? If Percy, Donald, Pelorus Jack, Nina, Dolly, Ono and the various 'Flippers' are anything to go by, it is quite possible they are. The Roman biographer Plutarch, who lived in the first century AD, summed it all up. He wrote:

> To the dolphin alone nature has given that which the best philosophers seek: friendship for no advantage. Though it has no need of help of any man, yet it is a genial friend to all, and has helped man.

We may find this a little far-fetched, and have already recognized that dolphins gain advantages during seemingly cooperative fishing ventures. They also actively encourage fishermen to part with a few morsels. At Bunbury, south of Perth in Western Australia, dolphins follow the fishermen, coming up alongside their boats for a free feed and sometimes diving underneath to spin the boats around in what seems to be play.

There seems little doubt, from the examples already quoted, that authentic two-way communication between man and dolphin has been achieved. But could we actually talk to them? Some scientists believe we can. One of them is John Lilly, who, apart from his work as a cetacean researcher, is a medical doctor interested in the brain. In 1960 he predicted:

> Within the next decade or two the human species will establish communication with another species: possibly extraterrestrial, more probably marine; but definitely highly intelligent, perhaps even intellectual.

Lilly thought it an optimistic prediction, but nevertheless was prepared to stick his neck out, and the animal he thought most likely to be in the vanguard of this two-way communication with man is the dolphin.

Big Brain

During the nineteenth century scientists began to dissect dolphin and whale brains and, comparing brain capacity with body size, came to the conclusion that the dolphin had the largest brain of any animal, including man, in proportion to its body size (the largest brain in the world is possessed by an adult bull sperm whale). Surely, some researchers thought, this must indicate great intelligence, maybe even an intelligence superior to our own?

It was an interesting thought, and the mystery and apparent charm of these animals most likely encouraged the belief that they have superhuman abilities. The almost complete lack of understanding of dolphin behaviour, coupled with their perpetual 'smiling' faces, seems to make them irresistible objects of human affection. It must have been, and perhaps still is, easy to interpret dolphin behaviour partly by relating it to human behaviour; but what may make a nice story does not necessarily make good science.

Anatomically, the dolphin brain *is* large, and the neocortex (the part with which we create, innovate and reason) covers 98 per cent of the surface (compared to 96 per cent in humans and 69 per cent in kangaroos). That seems impressive, until you learn that the dolphin neocortex is considerably thinner and less complex than that of the human brain.

Peter Morgane, at the Worcester Foundation for Experimental Biology, and Ilya Glezer, at the City of New York Medical School, together with researchers in the Soviet Union, have been able to map parts of the dolphin brain. Of particular interest has been the cerebral cortex. What the studies have revealed is that although the dolphin brain, during its evolution, became larger, it did not become more complex. The organization, revealed by Morgane and his colleagues, is more like that of its land ancestors that first went into the water about 50 million years ago.

Comparable land mammals, including man, show much greater changes. The human neocortex, for instance, consists of densely packed columns of neurones (nerve cells) that represent the processing units, interlaced with a complex microcircuitry, and within which are sandwiched areas of the brain associated with the sensory systems. It is this integrated system that facilitates memory and emotion, and is critical for learning.

The dolphin neocortex, by contrast, consists of fewer, but larger, columns of neurones linked by less complicated 'wiring'. The areas of the brain associated with touch, vision and hearing remain more generalized. Morgane has likened the structure of the dolphin brain to that of the hedgehog or the bat. Their brain structure evolved to something like its present state over 10 million years ago.

We must also consider the environment in which the brain has developed. Humans developed dexterity and complex tool-use, walked upright, lived socially, invented a complex language which has enabled them, according to some psychologists, to 'think' and generally change their environment to suit themselves. For that you need a substantial neocortex. The dolphin lives in the sea with other demands on its brain.

Dolphins are air-breathing mammals and as such must come to the surface at regular intervals in order to breathe. In addition to the normal cognitive and motor functions of other mammals, the dolphin brain must continually monitor what is up and what is down, how far away is the surface, when the air-supply is going to run out, how fast and how far it can travel in one breath, what the surface conditions are like in order that it can take a safe breath without being swamped in a heavy sea and so on. These actions are not all automatic. The brain must carry out a series of calculations in order to maximize all the activities that can be achieved with each lung-full of air.

In addition, the dolphin lives in a world in which sound dominates. The underwater explorer Jacques-Yves Cousteau was wrong when he called his book *The Silent Sea* and Coleridge was guilty of more than poetic licence when he wrote, 'We were the first that ever burst into that silent sea' in *The Rime of the Ancient Mariner*. The sea is a noisy place and many of the creatures that live in it contribute to the cacophony. The great whales have taken over the bottom registers, their very loud low-frequency calls travelling many kilometres across the oceans. The fish and lobsters are the trebles, with their mechanical clicking and clacking. And the dolphins have dominated the upper registers with clicks and burps that ascend into the ultrasonic, way above the normal hearing of humans.

That dolphins can hear ultrasonic sounds was proved in

tests on captive bottlenoses by Winthrop Kellogg, of Florida State University, in 1951. When sounds were played into the water, the dolphins changed their rate of swimming. They responded to sounds up to 80kHz – way above the upper limit of healthy human ears. At about the same time William Shevill and Barbara Lawrence, of the Woods Hole Oceanographic Institution and Harvard's Museum of Comparative Anatomy, worked with captive dolphins in a mud-bottomed Florida inlet. Their animals were trained to come to a feeding platform in response to pure tones. It was found that their hearing dropped off at 120kHz.

But dolphins not only appear to communicate in sound with burps and raspberries; they also interrogate their environment with sound. After observations made by Arthur McBride in 1947 and Winthrop Kellogg in the 1950s, research by Ken Norris, working at Marineland of the Pacific in 1960, showed that dolphins find their way about the sea by using echolocation: they 'see' with sound. Norris and his colleagues devised a way in which they could blindfold a dolphin with rubber suction cups and watch how it behaved.

Kathy was the subject, and she swam off across her tank as if nothing was amiss. She picked her way with ease through a maze of poles without touching one, and was able to locate small objects on the far side of her 10-metre tank. Small pieces of fish dropped next to a barrier in the maze would be scooped up without touching the obstacle. In another test she was invited to distinguish between a veterinary horse capsule filled with water and a piece of fish about the same size. Kathy took the fish every time. Norris had proved that dolphins used a sense other than sight with which to navigate and locate food. As dolphins have all but lost their sense of smell, and extrasensory perception has zero scientific credibility, sound was considered to be the likely candidate.

One of the problems, however, has been to understand how dolphins make these sounds. They don't have vocal chords and are rarely seen to blow bubbles when vocalizing. This is a controversial area of research. One school of thought has held that the sounds are produced in the larynx, as in other mammals, but echolocation clicks seem to emanate from the forehead rather than the throat. Indeed, if microphones are placed around the head of a 'clicking' dolphin the sounds can be triangulated to a location deep in the forehead, at the back of the nostrils. If probes are placed in the muscles of the larynx and nostrils it has been found that during sound production the larynx is quiescent while valves in the nostrils show muscle activity.

Ultrasound scans have given similar results. At Boston University, R Stuart Mackay and H M Liaw projected narrow beams of ultrasound at a frequency too high for the dolphins to hear and were able to identify the structures that moved during sound production. The apparatus was a modified foetal heart monitor of a type found in most modern maternity hospitals. The observers were able to observe the nasal plug and the vestibular, nasofrontal and premaxillary air sacs vibrate with the clicking of buzz sounds. Nasal diverticula (closed passages) on the right side vibrated all the time the clicks were produced, while the left nasal diverticula vibrated only some of the time. The vestibular sac inflated as the clicking sound was made, probably acting as a resonator. The nasal plugs were thought to be the site of the original sound production as air moved upward, presumably from the lungs. When the blowhole is closed, air recycles in the vestibular sac. Clicks seem to originate in the right diverticula. Whistles are thought to be generated in much the same way as human whistles except they are blown internally.

Dolphins have a pair of nostrils inside the head which come together under a single blowhole. The land ancestors of these animals probably possessed paired external nostrils, as most mammals do today; but as dolphins evolved into divers they needed a way of storing air if they were to make and use sound underwater. They developed a covering over the nostrils, the blowhole, with a complicated series of nasal sacs and valves below. Air is breathed in at the surface, and on diving, the blowhole is closed. The air trapped in the dolphin's respiratory system can then be passed across the valves in the nasal sacs to produce sounds, and then recycled through a complicated system of piping to be used over and over again as the dolphin echolocates underwater.

The click sounds are thought to leave the dolphin's body through the forehead. At the front of the forehead is a large fatty body known as the melon which focuses sound, much as a optical lens focuses light, about a metre in front of the animal's head. Donald Malins and Usha Varanasi, of Seattle University, have found a concentration of unusual lipids (fatty tissues) at the centre of the dolphin's forehead. The three-dimensional arrangement of these small molecules, which is rarely found in the lipids of other animals, suggests to the researchers that the area is indeed a 'sound lens'.

When the sound signal returns, having bounced off an object in the dolphin's path, it does not enter an ear canal but is picked up by the teeth in the lower jaw. The sound travels along more fatty material in the lower jaw, and thence is transferred directly to the middle ear. Sound production, transmission and reception is optimized for life underwater.

The rapidity with which clicks are emitted means that humans cannot hear individual packets of sound; rather, we hear trains of pulses much like – in Maura Mitchell's phrase – a creaking door hinge. One of the remarkable things about dolphin echolocation is the rapidity with which the ear and brain process the clicks. A click train may be made up of 700 units of sound per second. A dolphin is capable of mentally separating these units, listening to the individual echoes, and decoding the information while interrogating a target. In the human ear the sounds would fuse together in our minds at 20-30 clicks per second. No wonder dolphins have a substantial brain to process all this information.

The freshwater dolphins are considered relatively primitive. The sound beams they transmit are broad. Indeed, the blind Ganges river dolphin emits two broad beams of sound

– one that surveys the river bottom and another that monitors the way ahead. The dolphin travels along nodding its head, presumably to eliminate the blind spot between the two beams. It may be considered primitive in dolphin terms but, of course, it has become highly adapted to the murky environment in which the dolphin lives.

The bottlenose dolphin and some other coastal and open-ocean dolphins, on the other hand, have very sophisticated sound systems. Bill Evans, working in 1969 with the US Navy, revealed, with the help of a highly trained bottlenose dolphin named Scylla, who provided a train of clicks on command, that instead of the broad scanning beam, she could focus the sounds into a main pencil-thin beam of high frequency ultrasounds, alongside which she could form less intense beams that could be extended to the left or right.

Like freshwater dolphins, bottlenose dolphins have a melon in the forehead that acts as a sound lens. The sounds are focused into the main beam in front of the snout and, when turned on full, can be of such high intensity that, if they were any stronger, the sound waves would turn to heat.

In tests at the US Navy laboratory at Kaneohe Bay in Oahu (Hawaii), Whitlow Au and Earl Murchison discovered that a bottlenose dolphin could locate a tangerine-sized sphere at a distance of 100 metres. And it could interrogate an object with sound and discriminate between animate and inanimate, hollow or solid, and copper or wood. Bill Evans, some years previously, had a dolphin called Doris that could discriminate between two solid steel balls, one 2 inches and the other 2½ inches in diameter. Doris could also distinguish plates of glass, plastic, aluminium, copper and brass, and plates of different thicknesses.

There was, however, an unusual observation made by Evans when working with Scylla. It is reported in Forrest Wood's book *Marine Mammals and Man*, and quotes unpublished material. Scylla was swimming about when a live fish was put into her tank. As she chased after it she made the usual trains of echolocation clicks. Then she was blindfolded and another fish released. Curiously, she followed the path of the fish in silence – quite the opposite to what one would expect. Was she detecting very-low-frequency vibrations from the swimming fish? If so, it reveals the dolphin sound system to be even more remarkable. The dolphin is able, perhaps, to appreciate sounds right across the 'spectrum' from infrasounds to ultrasounds.

Even more interesting, as far as man-animal interaction is concerned, was that these tests involved the dolphin communicating with the experimenter. In an experiment in the late 1970s Murchison presented a female dolphin called Kae with various objects in her pen. He triggered the dolphin to scan her pen by playing a sound through an underwater loudspeaker. Kae scanned her pen and answered the tone for 'is there anything out there?' by either pressing a red ball for 'yes' or a blue ball for 'no'. If the answer was 'yes', Murchison used another tone for 'is it 'cylindrical? or 'is it round?' and so on. And then he asked, 'is it made of wood?' or 'is it made of steel?' Each time, Kae answered appropriately and would be rewarded with a Columbia river smelt. By a simple 'yes-no' system, Murchison could find out a great deal about the capabilities of the dolphin's sonar system. Even more significantly, the animal recognized such concepts as 'round' and 'cylindrical'. But the US Navy was not interested, at that stage, in communication or the subtleties of dolphin intelligence; it was simply keen to copy the dolphin's sonar system – and what an incredible system that is turning out to be.

Ken Norris has even speculated on whether the sound waves can look into the body of a fellow dolphin and evaluate its emotional state. Sound reflections, for example, from the beating heart of a fellow dolphin will tell the interrogator whether the dolphin is relaxed, tense or excited. With such a system of mutual examination, dolphins would have no need to say 'How are you?'; they would already know.

Even more extraordinary is the suggestion that these intense, high-energy sound beams are used by dolphins to stun or weaken prey.

With such a sophisticated sound system, it is inevitable that the dolphin should have a brain in which it can process and act upon the sound information it produces and the echoes it obtains. Its large brain, however, does not necessarily indicate an advanced intelligence. In fact, many sceptics suggest that the dolphin is no more intelligent than a dog, and the tricks performed at marine circuses are no more indicative of high intelligence than those that a dog, sea-lion or elephant performs in the circus tent.

Microbiologist Harry Jerison, of the University of California Medical School, has another theory. He suggests that the large neocortex devoted to the processing of sound could mean that the dolphin's echolocation system, like the human's visual system, is its 'reality'. A sound picture represents not only its perception of the outer world, but also its perception of self. Taking that a stage further, Jerison reminds us that dolphins are likely to be able to intercept each other's echoes. This has already been demonstrated in bats. If they can do that, then surely, suggests Jerison, dolphins will have developed not only an individual sense of self but also a communal one: a group consciousness.

Dolphinese

A large portion of the dolphin brain is undoubtedly geared to evaluate and react to information obtained by its echolocation system. There is also the prospect of dolphins being able to sample each other's signals. But, given all this, is there room in the dolphin brain for language? There have been, and there continue to be, those who believe that dolphins and their relatives have a language of their own, and that they are animals with whom we might eventually communicate and even have a meaningful conversation.

In a classic experiment in 1965, Jarvis Bastion, of the

University of California at Davis, placed a pair of bottlenose dolphins in adjacent tanks so that they were isolated visually, yet could still hear one another. The female was taught to push paddles in order to receive a reward. The male was given a similar set of paddles but received no training. Nevertheless, he pushed the correct paddle to obtain the reward. How did he learn which was the correct paddle? Did his mate next door tell him?

Throughout the tests the animals were heard to make whistles, squeaks, clicks and burps. At first it was thought that she was actually telling him which paddle to press. Later analyses of the results of the experiments cast doubts on this interpretation. There was, in fact, no conclusive proof that communication had taken place. That the male had somehow learned for himself the correct paddle to push, however, was a remarkable feat in itself.

Do wild dolphins have their own language? In the historical record, there have been many stories telling of dolphins and small whales using sophisticated communication systems. There are tales of dolphin schools avoiding boats attempting to capture or kill them. It was thought that somehow or other dolphins were able to tell each other: 'Steer clear of that boat – it's the same shape as the one that killed one of our school last week.' But that does not explain why thousands of dolphins are killed each year in the purse-seine nets of tuna fishermen, or why schools of pilot whales are driven inshore to be slaughtered on Faroes beaches and harbours. A sophisticated communication system did not save them. If it did exist, why did they not avoid the traps?

It is not clear whether any complex messages pass between dolphins, but it *is* clear that they are very noisy animals. There are squawks, whistles, squeaks, burps, groans, clicks, barks, rattles, chirps and moans. The bewildering array of sounds can be roughly divided into two types: pulsed and unpulsed sounds. The pulsed sounds include clicks and burst pulses. The bursts may be arranged into chuckles, chirps and click-trains. Trains of clicks sound like rusty hinges, while some burst pulses have been variously described as 'raspberries' or 'Bronx cheers'. The more continuous sounds are the whistles and squeaks.

It was once believed that the pulsed sounds were mainly for echolocation and the continuous tones for communication; but later research revealed that male bottlenose dolphins give pulsed 'yelps' during courtship and frightened dolphins give pulsed squeaks which may be alarm calls. John Lilly and Alice Miller revealed in 1962 that dolphins could click and whistle simultaneously. Of the pulsed clicks, they described soft clicks and hard-edged, very-high-frequency pulsed clicks, the latter associated, they thought, with echolocation.

Most research so far has concentrated on the whistling sounds; they are, after all, easier for humans to hear. And one thing is evident: dolphin and small whales that swim alone or in small groups use mainly click sounds, while those in large schools use clicks and whistles. There is evidently some link between whistling and living in a large group. Very-high-frequency clicks travel less far in water than lower-frequency whistles, so individuals in a large school would keep in better touch with their school-mates with the lower-frequency sounds.

One species of dolphin that has been studied in the wild by Ken Norris and his colleagues is the Pacific spinner dolphin. In bays around the Hawaiian islands, these dolphins spend the day in small groups, resting, playing and generally socializing. At night the groups come together and the large school goes hunting. For each activity of the spinner dolphin day, distinct sounds and sound patterns are used.

When resting the dolphins are relatively quiet, just making a few clicks. They swim slowly, with one half of the brain asleep while the other is alert. Gradually, during the afternoon, excitement rises and they begin to increase the number of whistles and burst pulses. The whistles are thought to represent individual identities, while the burst pulses indicate to each other their emotional states, whether they are angry or playful. It is as if each is greeting the other members of the school and asking if they are well and ready for the coming hunt. Each activity blends with the next. The sounds then change from predominantly burst pulses and few whistles to more whistles and fewer burst pulses.

Activity increases at the sea surface. Dolphins leap from the water and twist in the air. Underwater they roll over each other and emit sounds. The intensity of activity and noise gradually increases. Again, it is like a roll call, each animal indicating that it is 'present and correct'. Calling is not continuous but comes in short bursts separated by periods of silence. Each burst is thought to show how unified the group feels. If some do not join in the chorus, then it shows that the school is not yet ready to go. So the dolphins call and each time another member or two joins in, until all the dolphins have indicated their readiness to go hunting.

While the roll-call takes place and while the animals twist and turn in the air, they swim in a zig-zag fashion to and from the entrance to their bay. It is as if they are trying to decide whether they are ready to go to sea or not. At the peak of activity in the late afternoon they all head out in line abreast and, significantly, while hunting they use only clicks to scan the sea ahead.

Cape hunting dogs in Africa and other animals that hunt in groups, such as wolves and hyenas, behave in a similar way. Cooperation is the key to successful hunting. Before leaving, the animals twitter to each other, getting progressively more excited about the hunt and therefore more ready to hunt as a unit.

Dolphins in captivity, who have their meals provided, tend to whistle and squeak at meal times. Indeed, dolphins in situations that demand some kind of comment do, in fact, make distinctive sounds. Those riding bow-waves, for example, whistle away to each other. Similarly, if one is isolated from the school, the level of calling rises. Two schools meeting elicits a very noisy bout of whistling, squeaking and clicking.

There are suggestions that emotional meanings can be read into the sounds. Abrupt, loud sounds are given in aggressive situations. Sometimes these vocalizations are accompanied by non-vocal jaw-clapping or tail-slapping. What may be taken to be chuckles are heard during bouts of caressing and touching.

It is also believed that dolphins have their own individual identification sounds. In the spinner dolphin roll-call sessions, it is possible that each dolphin prefaces its overall 'I'm ready' call with a phrase representing its own 'name'. In an experiment in the Soviet Union in 1974, two neighbouring bottlenose dolphins, linked by sound only, both produced whistles and rhythmically related clicks. It was suggested that dolphins, like some songbirds, have an opening identification portion of the call (the whistle) followed by a more complex message portion (the clicks).

This birdsong parallel was continued further in the interpretation of the results of some experiments with tropical spotted dolphins. A male was captured and recorded. It produced, as all newly captured dolphins do, an almost continuous bout of whistling which the scientists interpreted as an alarm or distress call. The calls were then played back to the school from which the dolphin had been taken. They fled immediately. The same calls played to another school elicited curiosity but not flight. The captured dolphin's own school detected danger from the familiar, but frantic, call of the subject, whereas the 'strangers' were unable to appreciate the significance of the calls.

Sheri Lynn Gish, of the National Zoological Park in Washington, DC, previously worked at the University of California at Santa Cruz, where she had dolphins in separate tanks that were linked only acoustically. This, she thought, went some way to duplicating conditions in the wild when dolphins could hear but not see each other. Some patterns emerged from the results Gish obtained.

Dolphins in separate tanks did not overlap their calls and tended to match calls: a whistle was matched by a whistle from the dolphin in the other tank and a burst pulse was matched by a burst pulse. Gish also noticed a rhythm in the exchange of calls. Short calls passed rapidly between the two dolphins, while longer calls were greeted from the other tank by long periods of silence.

More recently, work by the cetacean researcher, Peter Tyack, at Woods Hole, has shown that most of the whistles heard from a school of dolphins are, indeed, signature whistles. The sender must use the call to tell the rest of the group that it is there. And research by James Ralston and Harvey Williams, of the Marine Mammal Laboratory at the University of Hawaii, has revealed that contained in their signature calls may be indications of their emotional state. They do not use extra calls that state emotions, but deliver their normal signature calls in a particular way.

Ralston and Williams analysed the calls from captive dolphins when swimming and socializing normally and when in stressed situations, such as being removed from the tank for a veterinary examination. The researchers discovered that under stress the configuration of the calls, such as pitch or duration, changed.

New research at Hawaii is using a sophisticated sound-monitoring contraption devised by Richard Ferraro, of the Institute of Applied Physiology and Medicine. A microcomputer, which can monitor every sound a dolphin makes, is fastened to its head by suction cups. The data to be analysed by the researchers will not only include the sounds themselves but will also indicate which dolphin is calling and what is going on at the time.

There is, then, no evidence of a complex dolphin language. There is, however, the staggering possibility – and it *is* only a possibility – that as dolphins rely so much on their acoustic sense, they may be able to project their thoughts (in the form of acoustic rather than visual imagery) on the mind of another dolphin in their school.

Wild speculation aside, though, the dolphin is proving to be a good mimic. Mimicry of social calls within a group, thinks Peter Tyack, is one way in which a group initiates social interaction and reinforces group cohesion. And mimicry within groups of a related species, such as the killer whale, is known to have produced distinct local dialects.

The evidence was demonstrated by Dean Fisher and John Ford at the University of British Columbia in Vancouver. The conversations between killer whales turned out to be rather dull. One of the first discoveries was that the sounds of any given killer whale pod are very stable. In a variety of situations the animals make the same associated sounds over and over again. In several different pods, distinct stereotyped calls were identified. These are exchanged when whales in a pod are spread out and foraging over a wide area. It is thought that these calls, like the signature calls of dolphins, ensure that all the whales stay in touch.

The dialects arose as a result of the isolation of one pod from another. Killer whale pods are essentially extended-family groups. Once an animal is born into a pod it rarely leaves the group. In this way a pod may grow up to 50 strong, although the average family consists of between six and 15 individuals. If an ancestral pod grows to a certain size, it is likely that it splits into two or more smaller groups, which spend progressively less time together. Initially, the calls given by such groups would be similar, but in time the dialect of each group would drift away into its own distinct sound and shape. In the Vancouver area, for example, there are resident pods that have their own clear patrol areas with their own distinctive sounds. Interestingly, further analysis of the sounds made by the present-day Vancouver pods has shown that several pods have common phrases in their calls. These ancestral groupings have been termed 'clans'.

Despite all the remarkable revelations, the Vancouver research team thought it unlikely that any highly structured and sophisticated message, like human speech, is being exchanged. More likely the information being communicated includes the position of the individual, its identity, emotional and activity state and pod identity through dialect. This did not, however, prevent a few people from trying to talk to killer whales. The methods used were, to say the least, bizarre, ranging from flute playing to keyboard composition.

Jim Nollman, who has given flute recitals to Mexican turkeys, accompanied howler monkeys in Panama and harmonized with timber wolves, grey whales and spotted dolphins, has played lute music to killer whales and even created a special underwater instrument. And the whales joined in, teaching Nollman new phrases which he could incorporate in his compositions and which he could use to guide dolphins away from the purse-seine nets of tuna fishermen through his dolphin rescue project, Interspecies Communication.

The most entertaining event, however, must have been psychologist Paul Spong's rock concert. Having failed to interest killer whales in the Johnstone Strait in musical recordings played through under water loudspeakers, he brought the rock band Fireweed to Alert Bay and gave the whales a free, live rock concert while sailing down the Strait. The whales were, like the local human inhabitants, not unexpectedly curious.

Erich Hoyt, meanwhile, tried to simulate killer whale calls on a keyboard synthesizer. In his book *Orca: the Whale Called Killer*, he describes how he slowed down recordings of whale calls and learned three killer whale phrases, each about two seconds long. On 20 July 1973 he was ready to have his first conversation. Several whales surfaced in the vicinity of the *Four Winds*, which was acting as a platform for a film crew trying to make the first major film about killer whales in the wild. The whales were leaping from the water and generally excited. Hoyt listened to the underwater activity via an underwater microphone and was startled to hear one of the phrases he had learned.

> With no notion of what to expect, I flipped on the tape recorder and turned up the volume pots on the synthesizer. I pressed the keys in the pattern I had devised, monitoring the

> imitation whale phrase as it passed out the underwater speaker. I held my breath. Two seconds went by. And then it came: A chorus of whales – three, maybe four – sang out a clear, perfect imitation of what I had just played to them – in harmony! They did not repeat their own sound; rather, they duplicated my human accent.

Hoyt was understandably over the moon; but there was more to come. After mimicking his synthesized phrase, they added a new one. It was almost an invitation to continue the conversation. He tried the rest of his repertoire, but the whales did not respond. The scientists who later heard the tape were intrigued. Had Hoyt inadvertently found the greeting phrase for that particular pod and had the whales realized that the sounds were coming from the boat and were man-made? Whatever the answer, it was again evident that members of the dolphin family are good imitators; perhaps even more significantly, these killer whales were not in a dolphinarium being coerced into mimicry with a reward – they were wild, free whales.

Although the more recent studies of dolphins and small whales are being carried out on wild populations, much of our present knowledge is based on research with captive animals. Paul Spong and Don White originally started their work for the University of British Columbia with a captive killer whale called Hyak in the Vancouver Aquarium. They tested the whale's visual acuity and discovered that above the surface it could see as well as a cat. Underwater, it relied mostly on sound.

It was this revelation that got Spong thinking about the world in which captive dolphins and killer whales are kept. The reverberations from the concrete sides, he thought, are more likely to silence the animals. They are deafened by their own sounds. After another incident, he began to question the value of keeping animals in captivity. One day, he was dangling his feet in the tank containing a female killer whale called Skana. The whale swam up to him, turned on her side, and rasped her teeth along the bottom of Spong's foot. Shocked, he jerked his foot away. Then, he put it back again and, sure enough, Skana, returned and did the same thing again. Spong was a little uneasy with Skana flashing her teeth at him but after a dozen more passes he plucked up enough courage to keep his foot in the water. At this point Skana lost interest. Spong suddenly realized that it had not been he who was studying the whale, the whale had been studying *his* behaviour. With an intelligence like that, thought Spong, the only place for a whale or dolphin is in the wild. He has spent much of his life since then campaigning that whales and dolphins should have that right.

But it was that same intelligence, brain size, ability to make a variety of sounds, and a seemingly special affinity for man that inspired scientists to choose the dolphin and some of its relations to figure in man-animal communication experiments.

Lilly and Friends

Aristotle recognized that dolphins made squeaks and moans, and suggested they might be able to mimic humans. He even went so far as to assert that they would be able to make vowel sounds with ease but might have difficulty with the consonants. In more recent times, John Lilly – who stuck his 'vulnerable neck out', as he put it, with his 1960s prediction – is the doyen of the 'smart dolphin movement'. As a student of the human brain he gained early notoriety for his isolation experiments in which humans, floating in water in darkness and silence, were deprived of visual, acoustic, tactile, pressure and gravity stimuli, and for his experiments with hallucinogenic drugs. Later he took up the cause of the 'intelligent' dolphin and initiated much of the pioneering work on dolphin-human communication. Lilly was convinced that dolphins are man's intellectual equal. He also considered that they might possess mystical, paranormal powers, and attributed intelligence to them. The object of his work was to discover a means by which dolphin and man could converse in order to understand dolphin language, culture, philosophy and even their system of ethics – a somewhat ambitious programme, to say the least.

Interestingly, while watching dolphins in tanks, Ken Norris noticed that bottlenose dolphins have 'echolocation manners'. He had wondered how animals possessing such a formidable weapon in their head did not accidently zap each other with their sonic transmissions. The answer, it turned out, is that they switch off their sound system when confronted with or when passing by another dolphin. So, if they have 'manners', why not more sophisticated social behaviour?

John Lilly wanted to find out, and he wanted the dolphins themselves to tell us exactly what they do and why they do it. He set out on this formidable task in 1955, while still working as chief of the Section on Cortical Integration at the National Institute of Mental Health at Bethesda, Maryland. At first his attention was focused on the dolphin brain. He saw his first one after a dolphin died from anaesthesia (it was not until a respirator was invented that could overcome the dolphin's voluntary breathing system that dolphins could be safely kept alive under anaesthesia), and Lilly was immediately impressed by its size and complexity.

Between 1955 and 1958 he visited the Marineland Research Laboratory, part of Marine Studios at St Augustine, Florida, the world's first oceanarium, and observed dolphin behaviour during training sessions. Already in those early days he was suggesting that dolphins are not animals on which you do experiments but that they are intelligent creatures with which you cooperate in scientific endeavour. This remark contrasted markedly, however, with the alleged treatment of an experimental animal at that time. Accounts of the same event differ.

In a brain-probing experiment, recalled by Lilly, a dolphin was kept immobile in a box through which passed running seawater. The Marineland staff were concerned for the welfare of the dolphin because the weather outside was very cold and the water passing through the box was cold. Lilly himself, writing in *Man and Dolphin,* recognized that bottlenose dolphins were most placid at '80° to 85°F', and that below 70°F a dolphin keeps warm by 'rushing around'. The experimental animal was restrained in water below 70°, and could not move its muscles sufficiently to maintain its body temperature. Reports of what happened next are at odds.

Lilly describes how the animal was returned to its holding tank and two other dolphins lifted it to the surface in order that it might breathe. They touched the area at the base of the tail which caused a reflex action of the tail muscles to thrust downwards and keep the head close to the surface. This observation was quoted by Lilly as evidence that dolphins are intelligent, caring creatures which help others who are in trouble. But, did it actually happen?

Marineland staff tell a different tale. In Forrest Wood's *Marine Mammals and Man,* the author, who worked at Marineland at the time, also recalls the event and states there were no other dolphins in the tank. The experimental dolphin, literally stiff from cold, was 'walked' by one of the staff until it could swim again of its own accord. Forrest Wood and Cliff Townsend (who gave the assistance) checked the record card for the day 16 January 1958. It read:

> Porpoise returned to experimental tank. Had to be supported and exercised for 30 minutes, then managed to swim alone. Much recovered few hours later.

Who does one believe? It is under this kind of shadow that one can understand how Lilly's work and results have often been greeted with scepticism by the scientific community. The criticisms were understandable. Lilly seemed to make exaggerated claims for events that under normal scientific scrutiny would be found to hold little water.

There is, for instance, the often-quoted example of communication between dolphins. In fact the story concerns those outsized dolphins, the killer whales. Lilly was at the University of Oslo, working at the time for the US Air Force, and gave a lecture about his suspicions that 'dolphins are more intelligent than we give them credit for'. In the discussion that followed the talk, an unnamed visiting scientist from the Whaling Institute told the story of 'several thousand' killer whales invading the area around a fishing fleet and taking all the fish, so that the boats caught nothing.

Nearby was a whaling fleet, from which the fishing fleet asked for help. Several whale-catchers or *shytter* boats were despatched and, after they had fired only one harpoon, the killer whales left the area within the hour. The story continued with the observation that fishing boats near whale-catching boats were free of killer whales while those without the benefit of *shytter* protection continued to be harassed by whales. The remarkable thing about this episode is that, according to the Whaling Institute scientist, the whale-catching boats and the fishing boats were identical – both were converted corvettes – except that the whale-catchers carried a harpoon gun on the bows. In *Man and Dolphin*, Lilly retells this uncorroborated story but quotes it as an example of dolphin intelligence and communication. The killer whales, he thought, recognized the harpoon gun as dangerous and told their fellow whales to avoid boats carrying such a weapon.

Can we accept such a story as evidence of intelligence and complex conversations on the part of these creatures? Who was the unnamed scientist? Had he witnessed the event or had he heard the story from somebody else? Had the process of 'Chinese whispers' distorted the tale from something insignificant to something fairly mind-boggling? Or was the tale a complete fabrication – the product of a young Norwegian whaler's imagination when far from civilization in the vast, cold, bleak Southern Ocean? It is in the light of such discrepancies and weaknesses in reporting that one must reconsider the validity of some of Lilly's work and, in particular, the conclusions he draws from it. The above story is just one of several quoted by scientists who have worked with dolphins, that questions the scientific value of Lilly's work. Nevertheless, there are a few interesting points to consider.

Lilly also recorded the sessions and realized that if the dolphin sounds were slowed down they resembled human speech. This influenced his choice of later experiments; and when in 1959 he was able to set up his own institute – the Communications Research Institute – on the island of St Thomas in the US Virgin Islands, Lilly first of all tried to teach the dolphins English.

In September 1961, one of his subjects, Elvar Dolphin, was able to make sounds that seemed to resemble those of human speech; but they were emitted at frequencies high above the human hearing range. Slowed down two to four times on the tape recorder, the sounds could be heard. Lilly likened the sounds to the babbling of a human baby. By 10 September Elvar had learned to say 'stop it' and 'bye-bye', and by 23 October he had added 'more Elvar' to his vocabulary. In December he said 'squirt water'. At this stage, Elvar was simply mimicking his human trainers. But did he know what he was saying?

In the 1960s expensive research facilities were constructed. A house was modified to include an office, complete with telephone, set in a shallow dolphin pool, with a balcony on the first floor built like a canal, along which the dolphins could swim. Margaret Howe, one of Lilly's assistants, lived 24 hours a day with a dolphin called Peter to 'develop mutual understanding and shared communcation'.

Within two months, Peter could manage to count up to 10 and almost produce some English-sounding words. The recordings are amusing, but mynah birds and budgerigars can do better, if not as quickly. Lilly, though, thought the dolphins were doing more than such a 'slavish job' and were, he contended, 'obviously doing a much more complicated series of actions'. The scientific community, however, was unconvinced, and in a review of *Man and Dolphin* in the magazine *Natural History*, researchers Margaret and William Tavolga described Lilly's work as 'scientifically unsound and naive'.

There were some aspects of the work, however, that merit further attention. The dolphins, in using English, were being asked to speak at the very bottom of their vocal range, and in the air with the blowhole closed. That is the equivalent, as Robert McNally put it, of us trying to talk underwater through the nose in a deep bass voice. Few people could do that but the dolphins did.

On one occasion, early in the experiments in 1963, the astronomer Carl Sagan went to visit Lilly. Elvar swam over to Sagan and, presenting his belly on the surface, invited Sagan to tickle him. Sagan obliged. Elvar moved away, then returned for a second bout, but this time stopped a few centimetres below the water. Again, Sagan obliged. This was repeated several times, each time the dolphin resting a little deeper than before. At last, the dolphin was too deep to reach and Sagan gave up. Instantly, the dolphin raised itself out of the water, stood on its tail, and, towering high above the two men, enunciated the word 'more'. Had the animal appreciated the actual meaning of the word 'more', or had it simply used the sound because it knew that this triggered the desired response from humans? Either way, it was a momentous occasion, both for Lilly and Sagan.

These preliminary experiments did turn up some useful information. Lilly's dolphins, like dolphins and killer whales in the wild, showed a remarkable ability for rapid mimicry. On some occasions, an animal made the sound even before the signal it was copying was finished. In another test, the trainer gave a series of vowel and consonant sounds and the dolphin was about 70 per cent accurate in copying not only the same number of sounds, but also the correct time intervals. Dolphins seem able to appreciate the quantitative aspects of human speech, if nothing else. English, however, was not the language with which to gain any significant insights into the dolphin way of life.

Lilly stopped his work on his first dolphin project in 1969 when five of his research animals, in his own words, 'committed suicide'. He became disenchanted with working with dolphins in captivity, preferring to see the animals in their natural home, the open sea. For many years he returned to his study of the human mind. But in the 1970s, realizing that little work had been done to take his own man-dolphin communication experiments any further, he began to take an interest in dolphins once again. This time he saw captive dolphins as living not in 'prisons' but in 'interspecies schools' where both dolphins and people learned about each other.

In 1975, Lilly returned to the research in a big way, forming the Human/Dolphin Foundation, funded by Hollywood society, which gave him a brand new hi-tech project. Dolphins and computers were kept at Redwood City, California, and the operational data centre was in the Malibu Hills in southern California, next to the Lilly home.

Lilly had taken account of the ape sign-language experiments and was particularly struck by the computer system devised by Duane Rumbaugh (see page 60). Several other dolphin researchers had also been impressed and wanted to know how to transfer the apes' system, which consisted mainly of a switchboard with 156 push-buttons that the apes could punch with fingers or fists, to one with which dolphins could push with their beak. Lilly thought this an odd way to proceed: dolphins do not have hands, like humans and apes, with which to manipulate things What they *do* have is a sophisticated sound system. Why not, thought Lilly, explore that route as a means of communication?

The project was given the acronym JANUS (Joint Analog Numeric Understanding System). It was equipped with state-of-the-art computing hardware and software. An attempt was to be made to bridge the gap between the dolphin's 2000 to 40,000 Hz vocal frequency range and man's 300 to 3000 Hz range. In addition, dolphins apparently 'speak' at a speed 15 times greater than we do. The computer generated dolphin-comprehensible wave forms which were matched to dolphin-viewable symbols on an underwater screen, while other computers analysed the dolphins' response.

What this meant was that a sophisticated computer-assisted sound system received and transmitted bleeps and whistles to and fro between captive dolphins and the human researchers. Dolphin sounds were matched to computer-generated sounds and to visual letters and other symbols which the dolphins could see on an underwater screen at the side of their tank. The dolphins reacted to the symbols on the screen, caused the sounds or symbols to change, or simply matched the artificial sounds with dolphin sounds. The object of the experiment was to create a new form of 'language' that was intelligible to both man and dolphin.

Tests started with 48 sound-and-symbol combinations, known as morphemes. Each sound-symbol combination was associated with an object, a place or an action. Lilly estimated that it would take about five years to work out a human-dolphin dictionary which could be used to communicate across the species boundary.

The two lucky 'students' to head the dolphin end of the experiment were Joe and Rosie; leading the human team was John Kert, a physicist from Hungary. At first the two dolphins learned simple tasks such as jumping and recognizing objects, and gained a 30-word vocabulary. In one experiment, the dolphins swam through a channel to touch a ball with a flipper, all under instruction via the hi-tech interface.

In a report in March 1983, Lilly claimed that one of the dolphins had actually spoken the words 'Throw me the ball'. Curiously, the words had a Hungarian accent. Had recordings of the physicist's own voice got mixed up in the equipment? Sceptics thought they had, and phrases like 'cetacean chic' and 'the dolphin as floating hobbit' began to appear in the local newspapers; as for published scientific results, little is available.

Flipper and Friends

In work started in the 1970s at the Flipper Sea School on Grassy Key, Florida, once host to the famous television dolphins, researchers adopted a less expensive approach to a dolphin-human communication system: they whistled. Indeed, they used a whistling language not unlike that used by shepherds working with trained sheep dogs.

They also invited dolphins to discriminate between pairs of objects. A picture of, say, a red-and-white spiral-shaped sea-shell was put on a board by the side of a dolphin's tank, and an assistant presented the dolphins with two real shells – one was a red-and-white spiral-shaped shell and the other a black-and-white cone shell. The dolphin unerringly pointed with its snout at the shell that correctly matched the picture.

In the Dolphinarium at Harderwijk, in the Netherlands, Wilhelm van Heel was working with a killer whale called Gudren. Van Heel played Gudren frequency-modulated tones (sounds that go up and down in frequency, like music) and the killer whale was expected to respond by touching relevant objects. This she learned to do with considerable ease, and if she was presented with an object accompanied by the wrong sound she would get very upset. Eventually, Gudren was able to demonstrate the remarkable ability of killer whales to mimic the sounds, and she also learned to associate them with the correct objects.

Van Heel then went one stage further: he introduced sound signals that represented verbs or action words, and Gudren was taught to recover objects on sound cues. The fascinating thing was that Gudren did not just accept what she was given: she began to talk back. During training sessions she could be heard to vocalize the 'action' sound signals that she had been taught.

Training dolphins to talk back was the object of experiments at Marineland of Florida, when William Langbauer attempted to teach dolphins a form of sign language, rather as the ape-language researchers had done with chimpanzees and gorillas. In the experiments, given the grandiose title 'The Porpoise Language Acquisition Project', Snoopy and his mate Betty were invited to match symbols. The trainers showed the dolphins a black-and-white shape on one board, and the dolphins were required to push the correct identical shape towards another magnetic board. After six months, they had progressed well and (as with the chimpanzee experiments) their vocabulary was gradually enlarged. Eventually, it is hoped the dolphins, like the chimps, will be able to take part in rudimentary 'conversations'.

Similarly, Dianna Reiss, of San Francisco State University, has been trying to unravel dolphin communication systems. The research, which has been going on for over eight years, is known as Project Circe and involves four dolphins at Marineworld Africa in Vallejo, California. The technique has been to identify and record sounds and behaviour by analysing video and audio tapes of training sessions. The Vallejo dolphins also have their own keyboard system. A dolphin can use its beak to push any one of nine keys. Each key has on it a symbol, and each symbol has an accompanying sound. When the dolphin pushes a key it hears a sound signal, and the action is recorded on a computer. The symbols represent objects the dolphins like to play with or eat, such as a ball or a fish.

When the experiments began, the dolphins quickly learned to associate the symbols with a reward, and the sounds generated with the visual symbols. And, in true dolphin tradition, were able to mimic the sounds generated and incorporate them in their calls. But Reiss, unlike some of her contemporaries, is realistic in recognizing what she might obtain from her experiments. She has likened her research to that carried out by scientists interested in the search for extraterrestrial intelligence and has presented papers at conferences organized by the International Astronomical Union, the Planetary Society, and the National Aeronautics and Space Administration (NASA). Basic to both dolphins and extraterrestrials, she feels, is the problem of detecting sound signals and recognizing any pattern in those signals when we are not clear about the nature of the signal in the first place.

Herman and Friends

Since the mid-1970s in Hawaii, Louis Herman, director of the Marine Mammal Research Laboratory, together with Douglas Richards and James Wolz, began another major project with two bottlenose dolphins. The dolphins learned to respond to visual and verbal signals in much the same way as sheepdogs; but there the similarity ended.

The background to Louis Herman's research includes many years of study of the dolphin's sensory systems, particularly its capacity to see and to hear, and of its learning and memory abilities. The dolphin, as Lilly had discovered, has a good auditory memory, being able to learn and remember long lists of sounds presented to it. Herman wanted to know if language – 'the most complex method known of information transfer from one creature to another' – could be added to the growing list of the dolphin's abilities. Instead of looking at the way a language might be produced by a dolphin, the research team concentrated on how an animal might understand language. To do this, Herman copied the techniques used by second-language teachers, who instruct pupils by getting them to carry out tasks. The level of understanding can be measured by how well the task is carried out.

This involved a completely new way of training dolphins. At a dolphinarium marine circus, the normal way to get an animal to learn new tricks is to entice it gradually with food (although Keith Laidler tells of the rather cruel technique employed by some circuses by which dolphins are taught to 'sing' by offering them food and taking it away at the last minute. This makes them sound-off with frustration. The sounding-off is then rewarded, and eventually the dolphins sound-off on cue). To teach a dolphin to go through a hoop, for example, the animal is presented with the hoop, rewarded initially for poking its head through, and then rewarded progressively as more and more of its body passes through until it swims through completely. A gesture or sound signal is provided to initiate the entire behaviour of swimming through the hoop.

Herman's technique differs in that the dolphin is first taught the word for 'hoop' and then the word for 'through'. The words can be combined into 'through hoop', and without special training the dolphin should know what to do. If a new word, such as 'gate' is introduced and linked to 'through', the dolphin should know to go through the gate as it did the hoop. In short, an object word and an action word are taught to the dolphin and combined with other words in new contexts; this gives the researchers flexibility in the training sessions.

Unlike the sign-language experiments with apes, when the gorillas and chimpanzees are invited to converse with the trainer, thereby producing 'words' which might be open to misinterpretation, Herman started out simply to examine the cognitive characteristics of dolphins with which he feels he will be able to make more objective interpretations of the results. Herman has been asking the questions: In what do dolphins specialize? What are their limitations? What are they good at? How do their successes and failures compare with those of other animals? What role does intellect play in the dolphin's world? In short, the main object of Herman's research has been not language but the creation of a tool for discovering a range of behavioural abilities of dolphins.

Herman moved to Hawaii in 1966, joining the University of Hawaii as an experimental psychologist. His first animal subject was borrowed from the local dolphinarium but, within a year, he had set up his own facility, gained a five-year research grant from the National Science Foundation, and was given two bottlenose dolphins by the US Navy.

Herman's pupils were Puka and Kea, who quickly learned a 12-word sonic vocabulary. They could understand 'fetch the ball', 'touch the ring' and so on, and could easily have learned a vocabulary of 50 words or more. But the experiment came to an abrupt end in 1977, when two former attendants sneaked into the pool, took the dolphins away and released them back into the wild. The animals, who had depended on man for their food for eight years, most probably died soon after release. Indeed, Herman saw one of the freed animals not long afterwards. Already she had been attacked and had one eye badly damaged. Unfortunately, she could not be recaptured: as part of her training she had been taught to avoid nets.

Despite this setback, Herman decided to carry on and in June 1978 two female dolphins, Akeakamai (Hawaiian for 'lover of wisdom') and Phoenix (after the reborn project) were caught in shallow waters off the mouth of the Mississippi and brought to Hawaii. Ake was taught a vocabulary of arm gestures, and Phoenix a set of acoustic sounds that were delivered as identifiable whistles and squeaks. Both the sound tones and the arm-waving signals represented objects and actions.

At first the two dolphins learned acoustic and optical 'words' for objects, such as frisbees and hoops, placed on the surface of the pool. If she understood, the dolphin had to touch the object with her nose and, if correct, she received an audible 'yes' signal followed by a fish as reward. An incorrect response was followed with a 'no' signal; but the dolphins became so angry when they got things wrong that this signal was eventually withdrawn from use. In their frustration, the dolphins would aim a frisbee or other object very accurately at the trainer. And so the accepted method for dealing with any misbehaviour after an incorrect answer was for the trainer to turn his or her back to the dolphins.

After seven months of initial training the dolphins had acquired a 20-word vocabulary, and were then introduced to simple word combinations. Herman wanted to determine whether the dolphins could take account of word order and syntax, so the two dolphins were taught different grammatical systems. For the task – 'collect the frisbee and take it to the basket', Phoenix learned a more usual English-style object-verb-object grammar. She responded to the acoustic sounds that literally said 'frisbee-fetch-basket'. Ake's gestural language took a more Germanic word-order with the verb at the end, so that she had to understand the entire sentence before acting. The same gestural signal would be 'basket-frisbee-fetch'.

The introduction of 'up' and 'down', 'top' and 'bottom', or 'gate' and 'hoop' began to make the commands more complicated. 'Phoenix-frisbee-fetch-gate', for example, meant that Phoenix was asked to fetch the frisbee and place it between two posts that served as a gate. She could also distinguish between left and right with commands such as 'fetch-right-pipe'.

More significantly, the two dolphins responded to the commands 'in' or 'through', such as in 'through-hoop'. And on one occasion, one was given the command 'person-through-hoop' to which she responded by pushing one of the trainers through the hoop.

Herman was meticulous in his experimental methods. In order to circumvent the 'Clever Hans syndrome' (see page 6), the trainers giving the arm-signal commands were blindfolded with opaque goggles. This avoided any misleading eye contact and inadvertent cueing on the part of the trainers. He acknowledged, however, that it is almost impossible to eliminate all human non-verbal 'body language' prompting. All the experiments, however, were videotaped so that third party observers can check the validity of the findings. And the findings are very interesting indeed.

For one thing, the dolphins not only associate 'words',

whether visual words or acoustic words, with actions and objects; they also seem to appreciate grammar. They successfully complete tasks, for instance, having received commands made of sentences they have not encountered before, and even understand sentences constructed with different grammatical rules. When presented with long, complicated sentences, such as 'frisbee-fetch-bottom-hoop' (which means in English 'take the frisbee to the hoop at the bottom of the tank'), Phoenix responded correctly when presented with the sentence for the first time. She ignored a ball on the surface and a hoop at the top of the tank, and took the frisbee to the hoop on the bottom.

In Ake's gestural language she was taught the concept 'in', and in response to the first-time command 'basket-hoop-in', she correctly collected the hoop and placed it in the basket. Indeed, in an assessment of tests during 1982, Herman was able to reveal that Phoenix got 85 per cent of 368 tests with familiar and novel word combinations correct, and Ake 83 per cent of 308 tests. Even the errors in the tests they got wrong were related to the 'indirect objects' of sentences or 'modifier' words such as 'top' and 'bottom'. They always understood the general sense of the word combination, even if they were not 100 per cent accurate in the interpretation. What the dolphins demonstrated was that they understood both the meanings of words and how word order affects meaning. This, according to some linguists and philosophers, is central to human language and a sign of intelligence.

Another ability was revealed in 'displacement' tests. A dolphin was asked to look for an object that was not present in the tank. After a slight delay the object was introduced along with an assortment of distraction objects. Invariably, the dolphin headed for the correct object and in doing so demonstrated that it understood not only a particular object but also the intrinsic attributes it possesses.

It is, perhaps, significant that it had never been the main purpose of these experiments to 'talk' to animals. The focus has been 'comprehension' rather than language acquisition and production. In this way, Herman's dolphin experiments have differed from the chimpanzee, orang-utan and gorilla projects. But you can't hold back a determined dolphin. Ake can tell her trainers whether something is present or absent from the tank by pressing 'yes-no' paddles. Telling about an object that is not there indicates some appreciation of symbolism. Ake must have had a mental image of the absent object in order to talk about it at all. And Ake has mimicked some of the computer-generated sounds and can label objects such as a ball, frisbee, hoop, pipe and person with her own vocal versions of the signals. The dolphins, despite Herman's original intentions, are talking back.

Ake and Phoenix are veterans of the experiments now, and they have been joined by a couple of bright young things: Elele, who is a fast learner, and her mate Hiapo. There is the idea that these dolphins might begin to talk to each other using the man-made vocabularies learned by Ake and Phoenix.

More recently Herman's dolphins have been introduced to the evils of television. Instead of presenting Phoenix with a real person to give her the arm-signals, Herman videotaped the trainers and showed them to the dolphin via a television monitor. Phoenix still received her reward, though. At the same time as the trainer on the TV monitor picked up a fish and threw it into the tank, a hidden trainer did the same thing 'live'. But that was only the start.

In the next stage of the experiment the videotaped trainer stood behind a black screen with holes cut in it, through which the arms could protrude. Wearing white gloves against the black background, the trainer gave the hand movements. In stage three, the trainer gave the gestures with closed fists; and in stage four, the trainer's hands were substituted by white balls on the end of black-painted sticks. All the dolphin saw was the movement of two white dots. Nevertheless, she performed the tasks well – getting things right 60-70 per cent of the time.

Interestingly, Herman tried out the white dot tests on his

laboratory staff. Those who got the tasks right scored between 20 and 85 per cent; those who had been with the experiment the longest did the best.

The results of these tests showed Herman that dolphins are capable of appreciating not only complex sound information but also visual information. This, he thought, might help in any future man-dolphin communication experiments. In fact, he is already taking that next step. His intention is to go beyond the simple 'yes-no' experiments and attempt to discover, in a similar way to the chimpanzee computer-language experiments at Yerkes, whether dolphins are capable of producing a man-made language.

The initial vocabulary consists of 50 symbols that can be displayed, 16 at a time, on a board on the bottom of the dolphins' tank. A dolphin is invited to select symbols by pressing its nose against the appropriate symbol on the board. Simultaneously, the selected symbol appears on a television screen which can be seen at the side of the tank. As the dolphin chooses more symbols they accumulate on the screen so the animal has a visual record of what symbols or 'words' it has chosen. In short, it can see what it has said.

Anything You Can Do....

So far, we have focused on the remarkable capabilities of the dolphin brain and on the affinities the dolphin appears to have for people. But it is beginning to be realized that dolphins are not unique in their learning abilities.

At Californian State University at Hayward, Ron Shusterman and Kathy Krieger have been duplicating Herman's dolphin experiments with Californian sea-lions. Using the arm-waving gestures, rather than sound signals, the sea-lions were taught a vocabulary of 20 'words' and could understand over 500 two-, three- and four-word combinations. They *do* however, have their limitations. They do not have such good memories as dolphins, and do badly in displacement tests. If asked about an object that is absent from the tank and which they have not seen in the past 10 to 12 seconds, they will answer incorrectly.

Despite these shortcomings, Shusterman believes that his sea-lions can understand information given in ordered sentences. This could mean that language is not as complicated as it has been made out to be. What the sea-lions seem to show is that the difference between animals and man is that animal behaviour is directly related to what is actually in front of them.

And while we're on the subject of seals and sea-lions, I mustn't forget Hoover. Hoover is a spotted seal that lives at the New England Aquarium on Boston's waterfront, but he is no ordinary seal, for Hoover can talk. Articulating his words in a clear Boston accent, Hoover says 'Hello, come over here', 'How are you', and when he is tired of his company he may round-off with 'Get out of here'. The aquarium staff claim no credit for his education. He appears to be acquiring his limited vocabulary from elsewhere. Unfortunately the 'talking seal' has been saying a little more than hello and goodbye. The aquarium is located in a not-too-auspicious part of town and, it seems, a local tramp has been stealing in at night and teaching the receptive seal a few indelicate expletives!

The Critics

What do the other animal-man communication experts think of the marine mammal work? As would be expected, there are those who are enthusiastic and who are impressed by the 'comprehension' approach adopted by researchers like Louis Herman. In the early days of the ape-language experiments it was thought that, if an animal reliably produces a symbol or 'word', it actually understands that word. What Herman and, more recently, the ape researchers have recognized is that training methods must emphasize comprehension skills as well as simple production skills.

Other critics are less impressed. Inadvertent cueing is always brought up, despite the lengths to which Herman and his colleagues have gone to reduce it. And dolphins, like apes, may be clever at realizing that, by playing the 'language game' with their trainers, they gain ample supplies of their favourite foods, heaps of praise and a very pleasant way of life indeed. All they are doing, suggests Herb Terrace, is learning behaviours that earn rewards, rather than learning a language as such. David Premack draws attention to the fact that human language does not consist only of objects and actions, such as those taught to dolphins and apes, but also of abstract concepts. At present these are, he claims, largely absent from man-animal conversations.

It would be interesting to carry out a series of sign-language tests with other animals, such as dogs, cats and rats, with which we have associated the possession of significant intelligence. Would they, like dolphins, apes and sea-lions, acquire an ability to appreciate vocabularies of symbols and 'sentence' combinations? And what about the seemingly 'intelligent' octopus? With a creature having so many limbs, a gestural sign language would take on a whole new dimension!

Realizing the Dream

On the west coast of Australia there is a group of wild dolphins which may give us an opportunity to realize Herb Terrace's original dream of teaching an animal some kind of system of communication that both it and we could understand and then getting it to tell us something about its friends out there in the wild. The dolphins live in Shark Bay, about 800km north of Perth, near the town of Denham.

On the east side of Peron Peninsula, which splits Shark Bay into two, a small number of bottlenose dolphins have become habituated to accepting food and will tolerate people touching and swimming with them. If one of these dolphins could be taught a sign language or something similar, then researchers might be able to interrogate a population of wild dolphins.

The centre of activity is a small caravan park called

Monkey Mia. The local inhabitants are familiar with Holey Fin and her offspring Nicky, Joy and Holly, Crooked Fin and her daughter Puck, Beautiful and her male baby Bibi, and his friend Snubnose.

The history of this group is obscure, but it is thought that the first dolphin to visit Monkey Mia was Old Charley. About 25 years ago, he herded herring beneath the wooden quay, making them easy for fishermen to catch. They threw him fish and he returned time and time again – always, incidentally, at the same time, 7.15am precisely. His departure was just as abrupt: at 8.30 he would scatter the fish shoals and leave, reappearing the next morning.

The fishermen also cleaned their catches on the quay and threw titbits to the dolphin. Other dolphins followed Charley's example and in 1964 a teenage girl, on holiday at Monkey Mia with her family, encouraged them to take fish directly from her hand. One of the first to do so was Old Speckled Belly, an ageing, almost toothless female who no doubt welcomed the free hand-out, as she would have had difficulty in holding-on to a slippery fish with so few teeth.

In 1975, the caravan camp was taken over by a retired engineer, Wilf Mason, and his wife Hazel. They were unaware of the dolphins before they purchased the property but were soon to discover how dolphins can almost take over your life. Soon after they arrived, they established a tradition of feeding the wild dolphins, and now up to a dozen animals come close to the beach every day; half of them take fish from the hand. Not only mature dolphins come. Some mother dolphins, such as Beauty and Holey Fin, have introduced their babies to the people standing in the shallows. Not all the dolphins come every day. Some stay away for several days – doubtless they are hunting at sea with the rest of the 100-strong school that frequents these waters – but they almost always return.

In 1982 the scientists arrived. They recognized that here they had a population of wild dolphins that had become habituated to the presence of people, just as Jane Goodall had established with the chimpanzees at Gombe and Dian Fossey with the gorillas in the Virunga Mountains. Richard Connor, from the University of Michigan at Ann Arbor, and Rachel Smolker, from the University of California at Santa Cruz, realized they had 'a window through which to view dolphin behaviour in a natural setting'.

There are about 150 recognized dolphins in the area. They can be told apart by the nicks and scars on their dorsal fins. Not all are habituated to the food handouts. Only eight or nine come to the beach where, apart from providing entertainment for visitors, they are being closely monitored by researchers. Their voices are recorded individually and each dolphin's behaviour is followed throughout the day. Hopefully, a picture will slowly emerge of dolphin social life and daily activity.

Already, the Monkey Mia dolphins have demonstrated a characteristic they share with humans: a partiality for sex for sex's sake. Most animals indulge in sexual activity almost automatically as a means of reproduction. Only in humans and some of the other higher primates does sex take on a social function. Male bottlenose dolphins constantly mount anything that passes, whether male or female. Males try to copulate with females, whether the latter are ready to mate or not. Two-day old males have erections. And, masturbation – until now associated with animals in captivity that are 'bored' – has been seen frequently in wild dolphin populations.

Studies like those at Monkey Mia have been adding more detail to the knowledge of bottlenose dolphin daily life that Randy Wells and his associates at the University of Santa Cruz have been gleaning from 20 years of study of wild dolphins at Sarasota and Tampa, Florida.

There is a common belief that dolphins are capable of altruism. Female dolphin mothers, for example, will form a 'playpen' in which their offspring can play, safe from predators. Babies stay close to and depend on their mother until at least their third birthday, and sometimes until their sixth, although 'babysitters' sometimes look after youngsters when the mothers go hunting.

One interesting observation is that after leaving their mother, dolphins segregate into male and female groups. Males may even pair up for long periods of time and long-term friendships develop; at Sarasota, two old-timers have been seen swimming together since 1975. On the other hand Richard Connor has seen gangs of male dolphins from one group intercept and capture a female from another group.

More disturbing reports come from some controversial research carried out by Philip Whiting, a professor of sociobiology at Harvard University who, like John Lilly, earned early notoriety for experiments with hallucinogenic drugs (only with chimpanzees rather than humans). The more recent work was with over 200 dolphins, some of which were at the Robertson-Hall dolphinarium at Los Angeles and the rest of which resided in a specially built tidal lagoon on the Californian coast. Whiting joined forces with marine mammal researcher Trefor Schmidt, and they set out to discover whether social animals show any racial prejudices. They chose to work with common dolphins.

Many of the experiments involved introducing 'strangers' to existing schools and watching what happened. Some dolphins, whether of the same species or not, were accepted, some were ignored and others were aggressively chased away. It is these observations of hostile encounters and, especially, the interpretations Whiting and his colleagues have put on them that has ruffled a few brows in the sociobiological community.

Common dolphins from Sri Lankan waters and from the Atlantic coast of Africa differ in colour pattern and shape from the schools in California. The African dolphins have a shorter snout and mottled undersides, and the Sri Lankan ones are darker skinned and have wider lips. When individuals from either of these groups were introduced to the Californian school they were rejected. Many were charged and butted, and some were assaulted by groups of Californian males.

In another part of the experiment, Californian dolphins were taken to Africa and introduced to schools there. The same thing happened. The Californian dolphins were chased away. On one occasion, a Californian dolphin was drowned by a group from the African school, each one taking it in turns to hold it underwater.

Back in California Schmidt used dyes to change the colour of Californian dolphins, and then reintroduced them to their own school. They were rejected.

In an article in *New Society*, Steve Platt reported an interview with Whiting in which he summarized his work. Whiting's conclusions were sensational:

> These experiments suggest that it is almost certain that dolphin behaviour is determined by genetically-determined responses to colour. In a very real sense it is possible to say that dolphins are colour prejudiced.

Not everyone agreed, and the debate continues to this day. What this work and the communication work *does* focus on, however, is some of the more sinister military and political motives for the research.

The Dream Under Threat

If we *could* talk – by sign language, computer-translator or whatever – with one of those wild dolphins and gain new insights into the natural world, to what purpose might we put that new-found knowledge? John Lilly voiced his deep concern in the introduction to *Lilly on Dolphins*:

> If and when interspecies contact is made, it may be used as a force for peace or as a further aid to warfare. It may be that we shall encounter ideas, philosophies, ways, and means not previously conceived by the minds of men. If this is the case, the present program of research will quickly pass from the

> domain of scientists to that of powerful men and institutions and hence somewhat beyond the control of the first venturers. When that time comes, I hope that the ideas here presented will help those men of good will to lead wisely and that they will be a bit better informed than they were in 1945 concerning another scientific advance, that time in applied nuclear physics.

The sad reality, however, is that Lilly's words have already gone unheeded. The men of war have already tuned into the world of the dolphin, and much of dolphin research today is funded by the US Navy. It started in 1960, when the US Navy began a project at the San Diego-based Naval Ocean Systems Center, in which a Pacific white-sided dolphin was studied by scientists and technologists who designed torpedoes. The dolphin's hydrodynamically efficient shape was clearly a useful model on which to base the future shapes of new underwater weapons and weapons-carrying submersibles. The way the dolphin uses its tail in a seemingly effortless way to speed through the water was also of interest.

The speed with which dolphins swim and the depths to which they can go were explored in some of the early experiments. It was discovered that bottlenose dolphins swim at about 16 knots and spotted dolphins at 21½ knots – not as fast as was expected. To keep up with ships, they hitch a ride on the bow-wave. Nevertheless, it was realized that to achieve even those speeds the dolphin's body needs to be hydrodynamically very efficient.

In diving tests, where free-swimming dolphins were invited to deactivate a pinger set deep in the sea, the dolphin was supreme. One male bottlenose dolphin called Lii managed to descend to 230m in 62 seconds. He eventually went down to 520m.

There was, however, a slightly more sinister side to the research. Submarine hull design was one thing, but the setting of explosive mines by kamikaze dolphins and the guarding of harbours by dolphin sentries armed with bayonets was quite another. Indeed, many stories were picked up by the popular press. Some were inevitably exaggerations, based on half-truths; and it is likely that the novel and movie *Day of the Dolphin*, written by Robert Merle, coloured a few minds.

The story is of a professor who works for the military and trains dolphins. By exploiting the bond between the male and female dolphin, he eventually teaches them to speak in English. The professor is then removed from the programme and the naval authorities take the dolphins to Vietnam, where one is sent to blow up a ship. The harness attached to the mine does not release but the female is able to bite the harness and free her companion. When they return to the submarine, they discover it has already sailed and realize that they were meant to be killed. Robert Merle ends his book with the dolphins confronting the professor. Just as they are about to leave, one says: 'Man is no good. We will stay in the sea.'

Sentimental and melodramatic, such stories are likely to pollute fact. There is, as a consequence, what might be termed a contemporary 'dolphin mythology'– a compound of genuine human concern and a bogus mysticism. The disturbing thing is that some of the stories of military research are not just sensational media fabrications – they are undoubtedly true. As soon as marine circuses began, it was realized that trained sea mammals could help people with their work below the sea. Consequently sea-lions, pilot and killer whales, belugas, Risso's dolphins, white-sided dolphins, and bottlenose dolphins have featured in underwater experiments.

Tuffy, for example, was a bottlenose dolphin that helped the aquanauts of Sea Lab II, a kind of underwater house in which three teams of 10 US Navy divers (including the NASA astronaut Scott Carpenter) lived at a depth of 60m for 15-day periods in 1965. The animal carried mail to and from the surface, swam to fetch tools, and even helped to guide divers back to the underwater habitat when the water was murky. When Tuffy was first brought to the project there was concern that as soon as this highly trained and expensive animal was released he would take advantage of his new-found freedom and swim away. In point of fact, he did just the opposite and stayed. In an account by one of the Sea Lab team, Tuffy's enthusiasm is clear:

> On the first day we took him to the base he swam inquisitively and somewhat excitedly to and fro between us, the equipment lying on the bottom and our quarters. He behaved as if he first wanted to make himself familiar with the terrain and to greet each one of us. Then he soon fitted into our team.... Our communications with him, principally by signs, soon worked excellently. He obviously enjoyed carrying our tools around for us.

This was one of the first times a relationship of this kind – between a trained but 'free' dolphin and people – had been witnessed. The animal was free to leave but researchers had relied on a piece of behaviour that was previously limited to man. Konrad Lorenz once described man as 'the inquisitive animal'. Unlike many other animals that lose the 'play instinct' at sexual maturity and are thereafter completely equipped to take up their niche in nature, man continues to find out about the world. Hans Hass has pointed out that the adult dolphin retains that innate inquisitiveness, too. This can be, and has been, exploited.

As Jacques-Yves Cousteau said, 'No sooner does man discover intelligence, then he tries to involve it in his own stupidity.' And the first hint that the military were seriously interested in sea mammals came with the release of several sequences of film by the US Navy showing a pilot whale called Morgan. Morgan was one of two-long-finned pilot whales, a killer whale, and a couple of sea-lions that were taking part in the Deep Operations Recovery System, a project in which sea mammals were trained to recover items such as torpedoes or tools from the bottom of the sea. The seals were first trained at the Navy Marine Bioscience Facility at Point Mugu in California. They learned how to support divers, swimming to and from the surface to depths of 180m carrying tools or messages in their mouth. The whales, first at the US Naval base in San Diego and then transferred to the US Naval Undersea Research and Development Center in Hawaii, went deeper – down to 500m. The programme was called Project Deeps Op and it included research with pilot and killer whales, and belugas (which could go deeper than the other two).

In the film, Morgan was seen to take a pincer-like steel grab in its mouth and fasten it to a sunken torpedo. It found the torpedo by locating a pinger, set in the torpedo's tip, that emitted a regular high-frequency sound signal that the whale could hear. As it attached the grab to the torpedo, the grab separated from the section held by Morgan and the whale returned to the surface. Balloons on the grab inflated and floated the torpedo to the surface.

So far, so good; fairly innocuous stuff – whale helps man. But in the early 1970s reports in such respectable newspapers as the *Wall Street Journal* and the *International Herald Tribune* caused considerable concern in the sea mammal research community. Activity centred on Cam Ranh bay in Vietnam. Some dolphins, the reports claimed, were being trained to guard harbours and warships; and what's more, they were said to be carrying special bayonets and syringes with CO_2 cartridges with which to kill underwater intruders. Others were supposed to have been trained to attach limpet mines to enemy ships.

The revelations were the result of Michael Greenwood's testimony to the Senate Select Committee on Intelligence in 1975. Greenwood had been with the sea mammal research programme from its early days and, during the hearings, he brought up the subject of the 'swimmer nullification' programme in which, he claimed, dolphins were equipped with the CO_2 cartridges (the sort issued to US Navy personnel as anti-shark devices) to kill enemy frogmen.

The US Navy was quick to scotch the rumours. One official responded to an enquiry from Hans Hass. 'Contrary to the sensational claims made in the press,' the statement went, 'no military use has taken place with explosive weapons, retractable daggers or similar weapons.' In another statement, to the *Wall Street Journal*, the official source questioned the sense in sending a $40,000 animal on a suicide mission or a mission in which the animal could be seriously injured when the same end could be achieved for far less money.

A year later a CBS television report stated that dolphins had, in fact, been trained for underwater sentry duties but did not attack the enemy frogmen. The trained dolphins simply detected enemy divers (which was particularly useful at night or in murky water), sounded an alarm, and forced the frogmen to the surface where they were captured. An interview with two of the US Navy team who trained with the dolphins appeared in the *International Herald Tribune*. During a training exercise, the Navy divers had acted as the enemy:

> When we tried to penetrate into the area, they discovered us immediately and brought us to the surface with their nose exactly to the correct place without any difficulty.... They rendered us so defenceless in practice that they could steer us in any direction they liked.

In an interview with Edward Linehan of *National Geographic* magazine in 1979, the director of research and development plans for the Chief of Naval Operations admitted to sending dolphins to Vietnam. They were there to test their portability over long distances and to see how they would operate surveillance and detection functions in the busy and murky waters of Cam Ranh Bay. They stayed for about a year, then spent another year in Guam before returning to the USA.

But the more ugly rumours would not go away. Hans Hass visited the United States and met some of the people involved in the dolphin training. When he returned to Europe he revealed that he had asked one diver if dolphins had been trained to kill enemy divers. The man said, 'Don't quote me by name, but they have.'

All went quiet for a while after the Vietnam war, but then in 1986 it was reported on the BBC World Service that the US Navy had requested $5.5 million dollars to fund the following year's secret programme for sea mammal training. The money was for the Advanced Marine Systems Project. In addition to the study of vision and navigation systems as part of a programme aimed at developing technology which will eventually imitate the underwater capabilities of animals such as dolphins, porpoises and seals, it was thought that these creatures were being trained for front-line military duties.

'Anonymous naval sources', according to the BBC report, described a test in 1985 in which dolphins were trained to detect mines laid in a South Carolina harbour. They found 80 per cent of the mines, which compared reasonably well with more conventional mine-detection methods. In the same report it was also revealed that the Soviet navy had been carrying out similar experiments with dolphins in the Black Sea.

That dolphins are being used for both overt and covert operations is now taken very much for granted, and it came as no surprise when in 1987 the Pentagon announced that Flipper and four other dolphins had been moved into the Persian Gulf to seek mines laid by the Iranians during the Iran-Iraq war. In Pentagonese it was said that the dolphins were there 'to provide surveillance and detection capability'. The dolphins apparently located mines, attached homing devices to them, and then returned to base, while US Navy ships went to blow them up. Flipper and his squad, though, were the same animals that had caused all the stir in Vietnam, and so speculation was rife that they were also underwater sentries, guarding against underwater saboteurs around the US Navy's floating command post in the Gulf.

At about the same time, more information became available. The budget for marine mammal research had reached $11 million a year and, since 1960, 240 animals had been trained for a variety of tasks. In the unclassified Project Quick Find, for example, sea-lions at the missile testing site at Cape Canaveral in Florida are trained to find spent rocket motors. They dive down to 200m, fix a line around the motors and then take the line back to the surface.

Dolphins, it was said, take part in a secret programme code-named 'Tag-a-Ship'. They learn to place homing-beacons on enemy submarines. It also transpires that trained navy sea mammals occasionally go 'absent without leave'. Tuffy, a sea-lion, became tired of war games and disappeared for three months. 'He had just got fat and goofed-off,' said a spokesman.

It also became clear that many animals were being enlisted into military service from wild stocks. The Marine Mammal Protection Act of 1972 in the United States places severe restrictions on the taking of sea mammals from the wild. Since 1986, however, the US Navy, with Congress approval, has been allowed to capture 25 animals a year, most being taken in Gulf coast waters.

In June 1988 the US Corps of Engineers let another cat out of the bag when they issued a public notice that they wanted to build underwater pens at the Bangor Naval Submarine Base, where 16 dolphins were to take up residence as underwater minders of the Trident missile nuclear submarines in Puget Sound (Washington state). Another facility would be at the Kings Bay Trident base in Georgia.

The following December, in the learned journal *Science,* David Morrison, the national security correspondent of the *National Journal,* wrote a major article on the sea mammal research programme of the US Navy. He discovered that in March 1987, the Chief of Naval Research had told the House Appropriations Committee, in a closed-door testimony, that dolphins would be used in 'expanding roles'. The naval chief also acknowledged that the Soviet navy were trying to catch up with US Navy military dolphin research.

In 1988 about 115 animals were working for the military and the projects had cost about $29 million between 1985 and 1989. The design aspect is still important with current research focusing on the dolphin's remarkable sonar system. Studies of the navigation and object-location capabilities of dolphins, particularly in poor visibility, are invaluable in designing new underwater gadgets for the military.

There was a scientific paper published in 1976, for example, that described an underwater experiment in which:

> Human divers, instrumented with 'bionic' sonar equipment based on the porpoise echolocation system and presented with targets earlier used in porpoise sonar discrimination tests, made scores as good as or better than the porpoises had.

The hydrodynamic research 'to determine if the dolphin does indeed possess a highly evolved drag-reducing system' also continues. This follows work that showed dolphin skin to possess three useful properties: it is porous, with a texture of 'wet velvet', it sloughs off easily, and it is very flexible. The sponginess absorbs water, and the mixture of skin cells and water reduces drag. In effect, the interface is like water sliding through water. Secondly, the tiny skin cells slough away in such a way that the animal literally swims out of its skin. And thirdly, the layer of skin, together with the blubber and muscles below, can change shape. It can conform its body to the optimal hydrodynamic shape for a particular swimming speed.

Also for the article, Morrison had interviewed Richard Trout, a trainer with Seaco, a company based in San Diego that trains the dolphins and sea-lions for the Navy. The trainer had previously told the press that the animals, when in the hands of inexperienced people, had been mistreated. Some dolphins, Trout claimed, had been struck when they did not carry out their tasks properly. Several dolphin deaths were also questioned. The animal welfare community and the US Navy, from their respective viewpoints, were both quick to respond.

An animal activist by the name of 'Charly Tuna of RainBoWarriors' gained access to the pens at San Diego and cut five of the nets. The dolphins stayed close-by, however, and returned to their pens the following morning. The Navy, for their part, announced that they had the lowest mortality record of any captive dolphin facility, and that the dolphins responded only to positive rewarding. Sam Ridgway, a veterinarian at the San Diego facility, wrote a rebuff to Morrison's article in *Science*. He drew attention to the fact that the animals perform their tasks while free. If they were maltreated, they could just leave. The few deaths had been natural, mainly caused by pneumonia, and there was no mystery or secrecy. The Marine Mammal Commission supported the Navy line. In a report issued at the beginning of 1989, it concluded that there was little evidence of the alleged mistreatment and that the US Navy does 'as good or better' a job of caring for its mammals as any other training facility.

The animal rights people were not convinced, and 15 separate groups took the US Navy to court in Seattle in April 1989 to stop bottlenose dolphins from being transported to the Trident base in Puget Sound. The water there is too cold for them, they argued.

The same month, the US Navy bought two Risso's dolphins from a fisheries cooperative at the port of Taijiin, Japan. The animals are known for their ability to dive deep – down to 1000m. They cost £1300 each, and were flown by US military transport plane from a Japanese airforce base near Nagoya to the Naval Systems Center in Hawaii. What they will be doing, only the military knows – and they're not saying.

5

Animals and Language

Language

In order for us to be able to talk to a member of another species, whether it be in the animal's own 'tongue' or in a system of signs that we have invented for the animal, we might need to consider whether any animals other than man communicate in what reasonably might be called a language, or have any appreciation of language. Do any of those series of squeaks, burps, whistles, screeches, twitters and grumbles, for example, constitute a language?

Some human groups have whistle, click and sign languages, so there is no real need for language to be defined as consisting essentially of complex words. Tone of voice, a sigh, a nervous cough or a scream of fright convey information that is important in communication; but these emotional expressions do not count as components of language. In fact, human language is paralleled by a non-verbal communication channel. The difference between the two is that you can lie with language but your non-verbal channel cannot lie. Actors can come close to 'lying' emotionally, but for the rest of us it is a more or less authentic reflection of our feelings, a barometer of emotions. 'I am a bright yellow banana' is a perfectly good sentence even though it is (probably) not true. A separation of meaning seems to be a good criterion of language. Animals in the main show the emotional, non-verbal system of communication, as they respond to events in their daily lives, but very few have what we might consider a language with which they can tell lies. Even humans mainly communicate in a non-verbal way, a phenomenon known as 'grooming talk' which has little to do with words but conveys the message – 'Please pay attention to me, I'm noticing you.'

Language can also report on emotional or bodily states but it has not necessarily evolved to facilitate those states. An awareness of self or consciousness, for example, does not require language.

In addition, language can tell of something that is not obviously visible, like a something or somebody on the other side of a hill or something in the past or in the future. It is clearly an advantage, in evolutionary terms, to communicate about something that is out of sight but not out of mind. Again, very few animals show signs of being able to do this. One that can is the chimpanzee.

Chimp Talk

A series of famous experiments were carried out with captive chimpanzees in a large chimp-breeding enclosure by E W Menzel. He would take two chimps out at a time, show them a pile of food under a bush, and then return them to the other chimps. All the chimps were then let out together. As they were youngsters they tended to stay in a tight bunch and, indeed, would not voluntarily go out at all unless they all went out together. The two chimps with the information about the hidden food, therefore, had to convince the others that there was good reason to cross the open enclosure. If the two scouts were dominant animals they would lead the others out. Subordinates would pull the fur of the others, look beseechingly towards the food, and if there was no response, they would throw a tantrum. In a similar test, the two chimps were shown two piles of food, one bigger than the other. The group went unerringly to the larger pile.

In another test Menzel introduced a snake instead of food – not a real snake, but a life-like rubber replica. One chimp was shown the snake placed, say, under a pile of leaves, and then he was returned to the others. Their behaviour when let out was even more intriguing. They were more cautious, creeping across the field with hair bristling and fear grins on their faces. They approached the pile of leaves and slapped it with their hands until the snake was uncovered. Then, they beat it to 'death'.

Menzel's chimps in these experiments had somehow told each other that there was food worth having or there was a snake of which to be careful. From the ape-language experiments we have seen that chimpanzees and other higher primates have the capacity for symbolic memory. Could it be that chimpanzees have reached a stage where they have the capacity to think ahead and make a mental image of something that is not readily visible?

Language, Thinking and Memory

If animals do not have language, then (according to some researchers) they also do not think. But might animals possess certain aspects of language and therefore have some capacity for thought? Expectation is important here. In many of the ape-language experiments, the abilities of the experimental animals have often been likened to the abilities of children. But, as Louis Herman points out, this comparison might not be valid. After all, when assessing children, we do not compare a child's swimming skills with that of a dolphin.

> A dolphin might think we don't demonstrate swimming until we can leap 15 feet from the water, stay under for 15 minutes, or swim at 20 knots. Otherwise we humans shouldn't really be calling it swimming.

And the same may be true of animal memory. Dolphins, for

example, are able to remember sounds or pictures presented to them several minutes earlier. 'So what?' say the animal behaviourists: some seed-eating birds can remember precisely where they have hidden caches of food for the winter; a digger wasp remembers the locations of each of its nesting chambers and maintains a running inventory on stocks of food in each hole; honey bees remember where the best flowers are located; and Arctic terns can remember the precise location of their nest site even though they have travelled half way around the world and back. The crucial difference between the dolphin and the birds, wasp and bee is that the latter group are unlikely to remember anything apart from what they are programmed to remember. They have no need to remember that which is of no survival value. They also seem to be able to remember only 'positive' things. The bee, for example, does not perform a negative dance to report the absence of nectar. Dolphins, on the other hand, can refer to things that are present and things that are absent. This flexibility in what a dolphin can remember in an experimental tank, however, is not something it obviously requires for survival in its natural environment. Dolphins have gone, as Herman put it, 'beyond what's necessary to what's arbitrary'. They cannot, though, appreciate the concept of 'not'.

Origins

The sign-language experiments have given us some clues about the language abilities of primates, although, as we have seen, these laboratory-based experiments have their limitations. The research has shown that captive apes can be taught a rudimentary linguistic form of signalling and can learn a large number of single words. They can learn the relationship between say, a plastic token and a sign. They can also relate to an object that they saw the previous day but that is not in front of them at present.

If, then, these apes in the laboratory can be taught to

perform this language-like behaviour, what do they say to each other in the wild? Somehow they must have evolved this ability. Natural selection must have favoured some sort of cognitive ability that has now become apparent to us, albeit in an unnatural setting. There are, as we have seen, problems. Herb Terrace has drawn attention to the fact that when an ape fails to perform a task – when it can't, for example, combine several words into a sentence or use a grammatical structure – we do not know whether that represents a lack of ability or a failure of motivation. An ape performing in the human world is asked to learn a human system of signalling. Using that system, we ask the chimpanzee questions in a very human-like way. This is like taking a Frenchman and asking him questions in Finnish. The subject is not going to give his best performance under these conditions. That was the reason that the thrust of Dorothy Cheney and Robert Seyfarth's work was not to teach the apes our language, but to tackle the apes and monkeys on their own terms.

Why, though, has human language evolved? Why is it that humans use this extremely complex form of signalling when other animals, as yet, do not appear to be capable of it? One way that natural selection could be considered to work is to say that 'necessity is the mother of invention'. We have this complicated form of communication because we need it. At some stage in the distant past an individual, by combining words into sentences, gained an advantage over the rest of the group, so that all the others had to learn it as well.

Man is clearly not alone in being able to use symbols to represent objects or events – witness the alarm calls of robins, ground squirrels and vervet monkeys. But is there a need for an animal to go any further? Possibly there is not. In many situations in which animals need to communicate they do not have the time to construct complicated signals, strung out in long series. Information transfer has to be rapid in the many life and death situations that arise in everyday life.

On the other hand, animals need not join their words end

to end. It could be that animals construct simultaneous, rather than sequential, sentences, with a number of components, each with an inherent meaning, combined as an overlay and broadcast together. We know, for instance, that American thrushes can sing two songs at the same time using the two halves of the syrinx independent of each other. This could be one very good reason why we have not recognized sentence construction in animals; and it would pose one heck of a problem if we were trying to interpret what exactly they were saying. Failing such a discovery, could it be that human language, as Naom Chomsky and others have suggested, is the Rubicon that divides humans from the other animals?

The studies, however, do give us some insights into the origins of our own utterances; after all, many of the primates being studied – vervet monkeys, baboons and chimpanzees – live in an environment similar to that thought to have been inhabited by early man a couple of million years or so ago. The clue is that these animals of the savannah are in tight social groups. Predation pressures are the probable reason why no solitary monkeys are to be found on the plains. Early man would likewise have found safety in groups; and to live in groups, individuals must communicate. Possibly the most important thing to a group member is another group member. And the most successful members of a group are going to be those most able to manipulate their fellows. As Robert Seyfarth once put it, 'group members become scheming Machiavellians in a cauldron of social competition', and the ability to use a vocal signal to manipulate others is going to have a strong premium placed on it; an ideal situation in which a sophisticated language could have evolved.

This leaves us with the tantalizing question with which we started: can we talk to the animals and, perhaps more interestingly, are we likely to get a reply? One problem Herb Terrace recognizes is that, although animals like the gorilla and chimpanzee can use sign languages, or speak in English, like an African grey parrot, they do not seem to show any *intent* to communicate. Alex the parrot (see Chapter 1) does not spontaneously call out 'Hey, I've just seen a green pen.' His motivation for mentioning the pen is the expectation of a reward.

Nevertheless, all the sign language experiments and the experiment involving a parrot learning English have demonstrated one important thing. Wild creatures are not automatons: they process information, make decisions, and demonstrate cognitive abilities which we had previously not appreciated. Aldous Huxley wrote in his essay, *The Education of Amphibia*:

> Without exception all languages are stupendous works of genius.... Without language we should merely be hairless chimpanzees.

Well, as we have seen, the 'hairy' chimpanzees, the feathered parrots and the 'wet-velvet' dolphins have begun to cast some doubts on the notion that 'linguistic communication and comprehension is the sole hallmark of thought and a unique human characteristic'.

Animal Einsteins

There is, however, one further scrap of evidence worthy of mention; it involves the ability to count – an ability already demonstrated by Alex the grey parrot, Peter the Dolphin – and now also by a chimpanzee.

It was once said that natural language is the language of sort and that mathematics is the language of size. Might other animals have the ability to appreciate and use this other numerical language too? The experiment that went some way to demonstrating that chimpanzees can count was carried out by Tetsuro Matsuzawa, of the Primate Research Institute at Kyoto University, Japan. A five-year-old female chimpanzee, Ai, was trained to use six Arabic numerals on a keyboard to name the number of items in a display. She also gave the appropriate colour and description, such as 'three red pens'. Ai is clearly a bright chimpanzee and she went on to show just how clever she is when she stole the key to her cage and made her escape with her mate Akira. To get out she had to open three separate locks. When recaptured Ai had the key in her mouth.

In another experiment at the University of Georgia, conducted by Roger Thomas, David Fowles and David Vickery, squirrel monkeys showed that they could appreciate the concepts of 'more' or 'less'. Rewards were hidden beneath white cards with circles drawn on them. Pairs of cards, one always having more circles than the other, were presented to the monkeys and they could gain a raisin by learning to tell the cards apart. Although this was not strictly a counting exercise, the monkeys were quick to tell a card with two circles from one with seven circles, and eventually went on to distinguish six circles from seven. The reward was always hidden below the card with lower numbers and so the researchers came to the conclusion that the monkeys had learned that 'fewer is correct'.

Interestingly, there was another example of an ability to count, but this time in a semi-wild situation. It involved cormorants and it was told to me by Miriam Rothschild. In the Far East, trained and tethered cormorants are used for fishing. Fishermen in small boats send the birds down to chase and catch the fish. When they return to the surface, a ring around the throat prevents the cormorant from swallowing the fish completely and it is disgorged to the waiting fisherman. In order to gain the birds' cooperation, the ring is removed after the seventh fish and the bird is allowed to swallow the eighth. If the fisherman loses count or forgets to give the cormorant the eighth fish, it refuses to dive.

Animal Interpreter

Clearly, the closer we observe animals, whether in captivity or in the wild, the more we come up with these kinds of surprises. But, in the human-animal communication business there is a long way to go before a truly meaningful conversation can be struck up. And there is still the problem of the work being taken seriously. Indeed, the scientific community has still to be convinced that this kind of research is respectable. The learned journal *Science*, for example, was reluctant to publish articles on artificial language in animals in the early 1980s, not long after Herb Terrace had voiced his criticisms. Louis Herman discovered this to his cost when trying to publish his first dolphin research paper on languge comprehension. In an interview with *Omni*, Herman recalled: 'It took us far too long to realize that policy and politics, not science, were at stake.'

However, there is, as far as I can ascertain, one instance of Herb Terrace's original dream, of being able to enlist the help of an animal interpreter, becoming a reality. It was not philosophically profound or even scientifically of great significance. But it was the first time that an animal using a man-made language 'told' us about the behaviour of animals in the wild.

Koko the gorilla was the translator and the event occurred one day when the chimpanzee researcher Jane Goodall visited Koko's California facility. Goodall was one of those pioneering researchers who, stimulated by the late anthropologist Louis Leakey, master-minded the famous study programme on wild chimpanzees in Tanzania, and founded the Gombe Stream Research Centre. Since 1965, Goodall had been studying wild, habituated chimpanzees and had gained remarkable insights into their everyday life. Confronted with a 'talking' gorilla, Goodall thought it an opportunity to ask a fundamental question. Talking through Penny Patterson, Goodall asked Koko whether he preferred his visitors to stand or to sit. Koko signed that he likes them to sit, and from that moment on, all observations of wild chimpanzees and gorillas have been carried out by sitting observers.

Whether Koko's suggestion is a personal preference or

represents the feelings of his fellow apes is not known, but what *is* known is that we are very far from being able to hold a conversation with another species in the wild. The obsession is still there, and scientists from all over the world will continue in their attempts to realize the goal of talking with animals.

But if we did communicate freely with other animals, how would we explain the way we have treated them? How would we explain to those monkeys and apes, incarcerated in the world's cosmetics laboratories, that they suffered for the good of man? How would we explain to the pilot whales whose annual migration takes them past the bays and beaches of the Faroes that they are butchered because it's an age-old tradition? How could we look an elephant in the eye and tell him that we killed his kin for the sake of a billiard ball, a piano key or an ivory necklace? What could you say to a young rhinoceros whose mother was slaughtered so that her horn could be turned into a dagger handle or form the basis of an ineffectual potion? How might you placate a large male crocodile who was destined to sire countless handbags, wallets and highly-priced luxury shoes? How would we justify our custodianship of the planet – the wholesale destruction of tropical rain forests and wetlands and the poisoning of the air and the seas?

The dolphins would certainly have much to say about our past attitude to them and their cousins the whales. Indeed, if they had access to the literature, I am sure they would take some interest in Lewis Radcliffe's paper 'Whales and Dolphins as Food', an *Economic Circular* from the US Fisheries Bureau in 1918. Contained therein were 22 recipes for the preparation of whale meat and 10 ways to cook up a porpoise. I wonder what our dolphin friends at Monkey Mia would make of that!

Bibliography

Pets and Their People, by Bruce Fogle, Collins, London, 1983

Living Wonders, by John Michell and Robert Rickard, Thames and Hudson, London, 1982

The Education of Koko by Francine Patterson and Eugene Linden, Andre Deutsch, London, 1982

The Talking Ape, by Keith Laidler, Collins, London, 1980

The Understanding of Animals, edited by Georgina Ferry, Blackwell and New Scientist, London, 1984

Animal Mind – Human Mind: Report of the Dahlem Workshop, edited by D R Griffin, Springer-Verlag, Berlin, Heidelberg & New York, 1982

The Ape's Reflexion, by Adrian Desmond, Blond and Briggs, London, 1979

The One Per Cent Advantage, by John and Mary Gribbin, Basil Blackwell, Oxford and New York, 1988

Almost Human, by R M Yerkes, Jonathan Cape, London, 1925

Animal Language, by Michael Bright, BBC Publications, London, 1984

Communication and Noncommunication by Cephalopods, by Martin Moynihan, Indiana University Press, Bloomington, 1985

Lilly on Dolphins, by John Lilly, Anchor Books, New York, 1975 (a revised edition in paperback including 'Man and Dolphins' 1961, and 'The Mind of the Dolphin' 1967, and several scientific papers by Lilly)

Conquest of the Underwater World, by Hans Hass, David and Charles, Newton Abbott and London, 1975

Orca: The Whale Called Killer, by Erich Hoyt, Camden House, Ontario, 1981 (1983)

The Dolphins' Gift, by Elizabeth Gawain, Whatever Publishing, California, 1981

Marine Mammals and Man: The Navy's Porpoises and Sea-Lions, by Forrest G Wood, Robert Luce, Washington and New York, 1973

The Magic of Dolphins, by Horace Dobbs, Lutterworth Press, Guildford, 1984

Index

Ai (chimp), 222, 223
Ake (dolphin), 196-9
alarm call, 83, 127-39, 177, 179, 220
alerting signal, 87, 88
Alex (parrot), 14, 15, 222
Ally (chimp), 72-3
American Sign Language (ASL), 12, 36, 39, 40, 41, 43, 45-7, 52, 54, 55, 65-7, 70, 73, 78, 79
animal artists, 57
ant, 1, 100-101
ape, 14, 16-78, 136, 190, 193, 195, 201, 202, 218-20, 225
Aristotle, 102, 140, 156, 184
Au, Whitlow, 173
Austin (chimp), 62-4

babbling, 37-8
Bastion, Jarvis, 175
bat, 1, 5, 169, 175
bat, fringe-lipped, 89
Beattie, Geoffrey, 11
bee, honey, 3, 79, 92-8, 219
beluga, 154-5, 208, 210
Berg, Judith Kay, 121
Berto (horse), 7
Betty (dolphin), 193
biological clock, 3
bird, 1, 80, 82-3, 86-8, 91, 105, 109, 111, 120, 124, 127, 135, 142, 179, 188, 219
blind trials, 39, 41, 62
body language, 1, 3, 9, 69, 109, 129, 161-2, 197
Booee (chimp), 43
Bonet, Juan Pablo, 34
brain, 13, 15, 21, 23, 65, 74, 92, 102-3, 152, 167-9, 172, 175, 177, 184-5, 200
Bronx cheer, 30, 177
Brown, Roger, 40
Bruno (chimp), 43
Buissac, Paul, 12-13
butterfly, large blue, 100-101

call, 80-83, 85-9, 92, 121, 123, 127, 129-39, 177-81, 193
cat, 2-4, 10, 13, 30, 39, 56, 69, 202
cephalopod, 102-5
Charkovsky, Igor, 149
Charles-Michel, Abbe, 34
Cheney, Dorothy, 130-38, 220
chimpanzee, 13, 16-44, 45, 49, 50, 57, 59-60, 62-6, 69, 71-5, 77-8, 193-5, 199-200, 217-18, 220-22, 224
chimpanzee, pygmy, 17-19
Chomsky, Naom, 65, 74, 221
chunking, 76
clan, 181
Clever Hans (horse), 5-8, 11, 39, 73, 78, 197
click, 170-72, 176-9, 216
Clutton-Brock, Tim, 89-90
cod, 106
Cody (orang-utan), 26-32
Cohn, Ron, 52, 55-6
colour change, 102-3
computer, 5-7, 59-64, 74, 78-9, 105, 190-91, 193, 199-200, 206
Coolidge, Harold, 17
cooperative fishing, 141-5, 167, 223
cormorant, 223
counting, 5-8, 15, 188, 222-3
courtship, 82, 88-9, 106-9, 177

cueing, 6-7, 13, 39, 54, 133, 136, 197, 201
cuttlefish, 102, 104

dance, 92-8, 219
deer, red, 89
Desmond, Adrian, 72
dialects, 18-20
Dionysus, 156
display, 15, 36, 104-6, 114, 128
Dobbins, D, 72
Dobbs, Horace, 146, 163-4
dog, 2, 3, 5-6, 8-10, 13-14, 39, 128-9, 133, 150, 175, 178, 192, 195, 202
dolphin, 1, 13-14, 84, 114, 118, 140-215, 218-19, 222, 224-5
dolphin, Atlantic humpback, 142
dolphin, bottlenose, 142, 159, 170, 173, 176-7, 179, 185, 194, 196, 202, 204, 207-8, 215
dolphin, common, 205
dolphin, contact with, 149-67
dolphin, Pacific spinner, 177, 179
dolphin, rescued by, 145-8
dolphin, Risso's, 157, 208, 215
dolphin, river, 143-5, 172-3
dolphin, spotted, 179, 182, 207
dolphin, white-sided, 207-8
Donald (dolphin), 158-64
Dorner, D, 74

echolocation, 143, 170-72, 175, 177, 185, 214
electricity, 107-109
Elele (dolphin), 199
elephant, 1, 114, 119-27, 175, 225
Elvar (dolphin), 188
Evans, Bill, 173
extraterrestrial, 193-4

Fellow, 10-11
Ferraro, Richard, 180
firefly, 1, 98-100
fish, 106-9, 141-5, 169, 193, 225
Flipper (dolphin), 151, 155, 166, 192, 212
Ford, John, 188-9
Fossey, Dian, 44, 204
Fouts, Roger, 41-3, 73, 76
Fowles, David, 223
frog, mud-puddle, 88-9
Fungie (dolphin), 165
Furness, William, 24-5, 27, 29

Gardner, Allen and Beatrice, 35-41, 43-5, 50, 72-3, 76
Garner, R L , 20
Gawain, Elizabeth, 148, 152
Gibson, Robert, 90
Gish, Sheri Lynn, 179
Glezer, Ilya, 168
Goodall, Jane, 35, 203, 224
gorilla, 13, 17-20, 24, 27, 33, 44-57, 65, 71-3, 78, 193, 195, 199, 204, 221, 224
Gould, James and Carol, 97
grammar, 12, 54-5, 58, 60-62, 69-70, 74, 104, 197-8, 220
Gribbin, John and Mary, 17, 18
ground squirrel, Belding's, 128-9, 220
Gua (chimp), 24, 35
Gudren (killer whale), 192
Guinness, Fiona, 90

haddock, 106
Hamilton, G R, 117
Hanlon, Robert, 103
Hanschen (horse), 7
Hass, Hans, 147, 209, 211
Hauser, Mark, 135
Hawkings, Anthony, 106
Hayes, Keith and Catherine, 25
Hediger, Heini, 11
Heel, Wilhem van, 192
Heffner, Rickye and Henry, 121
Herman, Louis, 84, 194-201, 218, 224
Hess, Eckhard, 4
Hiapo (dolphin), 199
Hillier, Barbara, 52, 55
honey guide, 79-80
Hoof, Jan and Anton van, 36
Hoover (seal), 201
horse, 5-8, 159
Hoyt, A M, 24
Hoyt, Erich, 182-3
Humphrey (whale), 83-5
hunting, 33, 49, 80-82, 178

Imgraguen, 142-3
imitation, 30, 34, 38, 40, 51, 67-9, 81-2, 183
infrasound, 120, 122-3, 125-6, 174
intelligence, 5, 11, 16, 19, 21, 24, 27,

102, 136, 162, 167, 174-5, 184-5, 187, 193, 198, 202, 209
invention, 38, 47, 58, 71, 72
Ioni (chimp), 23, 35

JANUS, 190-92
Jerison, Harry, 175
Johanson, Don, 19
Jozsef, Stephen, 152

Kathy (dolphin), 170
Kea (dolphin), 196
Keller, Helen, 22-3
Kellog, Winthrop, 24-5, 170
Kert, John, 191
keyboard, 8, 45, 56-7, 60-65, 182, 193, 222
Koko (gorilla), 44-58, 61, 64-5, 72, 224
Kortlaandt, Adrian, 36
Krall, Karl, 7-8
Krebs, John, 86-7
Krieger, Kathy, 200

Ladygin-Kohts, Mrs, 23, 25
Lady Wonder (horse), 8
Laidler, Keith, 26-32, 194
Lamb, F Bruce, 143
Lana (chimp), 60-64
Langbauer, William, 120, 193
language, 11-12, 14-15, 20-23, 25, 32-4, 40, 43-5, 49, 57-60, 64-5, 69-74, 76-7, 79, 82, 85, 92, 98-9, 101-5, 108, 129, 137, 169, 175-6, 180, 185, 191-5, 198, 200-202, 216-25
Lawrence, Barbara, 170
learning, 50, 57, 59-61, 66-8, 73-6, 105, 131, 135, 150, 160, 168, 192-4, 196, 200, 202, 219
Leon, Pedro Ponce de, 34
lexigram, 60, 62-4
Liaw, H M, 171
Lii (dolphin), 207
Lilly, John, 13-14, 154, 166, 177, 184-91, 194, 205-7
Lissman, Walter, 107
Lloyd, James, 99-100
Lorenz, Konrad, 209
Loulis (chimp), 43
Lucy (chimp), 41-3, 77
Lucy (prehistory), 19

McBrearty, Dennis, 165
McBride, Arthur, 170
MacKay, R Stuart, 171
McNally, Robert, 189
McVay, Scott, 111
Malins, Donald, 172
man-dolphin communication, 153, 167, 184, 190, 192, 200
Matsuzawa, Tetsuro, 222
Meehan, Joseph, 6
Menzel, E W, 217-8
Mettrie, Julien Offray de La, 33-4
Michael (gorilla), 52, 55-6
Midgeley, Mary, 77
military, 206-15
Miller, Alice, 177
mimicry, 14, 71, 80-81, 180-84, 188-9, 192-3
Mitchell, Maura, 159-64, 172
modifier, 105, 198
molecular clock, 18
monkey, 14, 21, 33-4, 41, 129-39, 182, 220-21, 223, 225
Monkey Mia, 203, 225
monkey, squirrel, 223
monkey, vervet, 220-21
Morgan (pilot whale), 209-10
Morgane, Peter, 168
Morrison, David, 213-4
Moses (chimp), 20
Moss, Cynthia, 122-3
motherese, 8-10
moulding, 29-30, 37, 44
Moynihan, Martin, 102, 104-5
Muhamed (horse), 7
Munn, Charles, 15
Murchison, Earl, 173-4

Nathanson, David, 150-51
Nelson, Don, 109
nightmare, 49, 51
Nim (chimp), 65-78
Nollman, Jim, 182
Norris, Ken, 170, 174, 177, 185
noun, 53, 58, 105

open words, 40-41, 58
Opononi Jack (dolphin), 158
orang-utan, 18-19, 24-32, 77, 199
Orne, martin, 78
Osten, Wilhelm von, 5-7

parrot, African grey, 14-15, 222
Patterson, B, 117-18
Patterson, F (Penny), 45-56, 64-5, 76, 224
Payne, Katherine, 111, 113, 120-21, 124, 126-7
Payne, Roger, 111-12, 118
Pelorus Jack (dolphin), 157
Pepperberg, Irene, 14-15
Percy (dolphin), 164
pet, 2, 3, 8-10, 69
Peter (chimp), 21-3, 35
Peter (dolphin), 188, 222
Pettito, Laura, 70
Pfungst, Oskar, 6-7
Phoenix (dolphin), 196-9
phrases, 38, 54, 111-13, 182-3
pigeon, 58, 75-6, 120
pigment cells, 103-4
pivot words, 40, 58, 67
playback, 80, 83, 85-92, 125-6, 131-2, 134-5, 137, 179
Pliny the Elder, 140-41, 156
Pliny the Younger, 156-7
Poole, Joyce, 122-3
Premack, David and Ann, 57, 59, 73-4, 77, 202
Pribram, Karl, 45
primate, 14-15, 24, 31, 49, 57, 64, 91, 105, 129, 130, 136, 138, 218-9, 221
Prince Chim (pygmy chimp), 16-18
protowords, 127
Puka (dolphin), 196

Ralston, James, 180
Rand, Stanley, 88
rat, 58, 75, 202
Reid, Howard, 81
Reiss, Dianna, 193
Rhine, J B, 8
Richards, Douglas, 87, 194
Rickard, Robert, 8
Ridgway, Sam, 215
Ristau, C A, 72
roaring, 89-91
Robinson, John, 91
Robinson, Scott, 128-9
Romanes, George, 19
Rothschild, Miriam, 223
round dance, 83
Rumbaugh, Duane, 60, 62, 190
Ryan, Michael, 88
Ryle, G, 71

Sarah (chimp), 57-61, 75
Sarich, Vincent, 18
Savage-Rumbaugh, Sue, 62
Savage, Thomas, 44
Schmidt, Trevor, 206
Schreiber, O W, 110
Sea Lab II, 208-9
sea-lion, 3, 175, 200-202, 208-9, 213-14
Seidenberg, Mark, 70
self, 39, 59, 67, 175, 217
sentence, 20, 57, 59-62, 65-70, 74, 197, 198, 201, 202, 220-21
Serpell, James, 15
Seyfarth, Robert, 130-38, 220-21
shark, 5, 58, 109, 145-8, 210
Sherman (chimp), 62-4
Sherman, Paul, 128
Shevill, William, 110, 114, 117, 119, 170
Shusterman, Ron, 200-201
Sicard, Abbe, 35
signal, 34, 38, 87-8, 101, 104, 117-18, 120, 122, 127, 130-31, 134, 137, 192-7, 199-200, 210, 219-21
signature calls, 87, 180-81
sign language, 7, 23, 26, 33-7, 42-3, 46, 49, 57, 65, 68, 70, 73-4, 76-7, 99, 130, 138-9, 162, 190, 193, 195, 202, 206, 216, 219, 221-2
signs and signing, 12, 22, 29, 34-6, 73, 77, 78, 99, 127, 209, 216, 219, 224
Skana (killer whale), 183-4
Skinner, B F, 27-8, 37-44, 46-54, 56, 65-7, 69-72
Slater, Peter, 82
Smith, Betsy, 151
Snoopy (dolphin), 193
song, 80, 82, 86-8, 91-2, 110-117, 124, 221
Soviet Navy, 212, 214
speech, 23, 26-7, 29, 53, 57, 65, 69, 71, 150, 181, 188-9, 208, 221
spishing, 82-3
Spong, Paul, 182-3
squid, 102, 104-5

stonechat, 127-8
Strusacker, Tom, 130
Sullivan, Anne Mansfield, 22
Symbol, 22-3, 58-64, 67, 73, 76, 191, 193, 199-202, 218, 220
syntax, 40, 65, 70, 74, 104, 197
synthesizer, 57, 64-5, 182

Tavolga, William and Margaret, 189
Temerlin, Maurice and Jane, 42, 77
Terrace, Herb, 1, 12, 66-78, 75-8, 139, 202, 220, 221, 224
territory, 82, 86, 91-2, 108
testing, 39, 54, 61-4, 66, 68, 121, 166, 183, 191, 198-9
thinking, 12, 43, 49, 74-6, 97-8, 169, 218
Thomas, Elizabeth, 120
Thomas, Jeremy, 100
Thomas, Roger, 223
Tick-tack, 34
tit, great, 86
titi, 91-2
token, 57-8, 61, 78, 219
Toto (gorilla), 24
towhee, rufous-sided, 87-8
Trout, Richard, 214
Tuffy (dolphin), 208-9, (sea-lion), 213
Tuttle, Merlin, 89
Tyack, Peter, 86, 114-15, 180-81

ultrasound, 122, 169, 171, 174
US Navy, 196, 207, 209-15

Varanasi, Usha, 172
verb, 40, 53, 58, 67, 105
Vickery, David, 223
Viki (chimp), 25-6, 29, 31, 35
vocabulary, 14, 24, 36, 38, 42-3, 46, 51, 53, 55, 58, 60, 63, 68, 70, 102, 137, 188, 191, 193, 196-7, 199-202
vocal communication, 14, 24-5, 31, 47, 57, 64-5, 81-2, 84, 89, 102, 106, 137-8, 192, 221

waggle dance, 93-4
Washoe (chimp), 26, 35-41, 43-5, 47, 49-51, 53, 55, 57-9, 61, 66, 69, 71
Watkins, William, 119
Watlington, Frank, 111
Webb, Douglas, 118
Wells, Randy, 204
whale, 1, 13, 109-19, 155, 167, 169, 176, 177, 181-4, 187, 210, 225
whale, fin, 117-18
whale, humpback, 83-6, 110-18, 124, 143
whale, killer, 84, 118, 143, 181-4, 187, 189, 192, 208-10
whale, pilot, 208-10, 225
whale, sperm, 119, 167
whistle, 2, 82-3, 86-8, 128, 171, 176-80, 191-2, 196, 216
White, Don, 183
White, Terry, 19
Whitehead, Hal, 117
Whiting, Philip, 205
Wiley, Haven, 87
Williams, Harvey, 180
Wing, Stephen, 99-100
Winn, Lois and Howard, 110-11
Witmer, Lightner, 21-3
wolf, 1, 9-10, 178, 182
Wolz, James, 194
Wood, Forrest G, 173
Woodford, Jack, 8
Woodruff, Guy, 75
word, 12, 14, 20-21, 23-5, 28, 30-31, 33, 35, 38, 40, 42, 46-8, 50, 54, 57-62, 65, 67-9, 71-4, 76, 105, 129, 131, 138, 150, 195-8, 200-201, 216, 219-20
writing, 21-2, 57-9

Yerkes, Robert, 16, 25, 35
yerkish, 60-62, 78

Zarif (horse), 7
Zihlman, Adrienne, 18-19